CHEMICAL INDUSTRIES

A Series of Reference Books and Text Books

Consulting Editor
HEINZ HEINEMANN
Heinz Heinemann, Inc.,
Berkeley, California

Volume 1: Fluid Catalytic Cracking with Zeolite Catalysts, *Paul B. Venuto and E. Thomas Habib, Jr.*

Volume 2: Ethylene: Keystone to the Petrochemical Industry, *Ludwig Kniel, Olaf Winter, and Karl Stork*

Volume 3: The Chemistry and Technology of Petroleum, *James G. Speight*

Volume 4: The Desulfurization of Heavy Oils and Residua, *James G. Speight*

Volume 5: Catalysis of Organic Reactions, *edited by William R. Moser*

Volume 6: Acetylene-Based Chemicals from Coal and Other Natural Resources, *Robert J. Tedeschi*

Additional Volumes in Preparation

ACETYLENE-BASED CHEMICALS FROM COAL AND OTHER NATURAL RESOURCES

ACETYLENE-BASED CHEMICALS FROM COAL AND OTHER NATURAL RESOURCES

ROBERT J. TEDESCHI
Tedeschi and Associates, Inc.
Industrial Consultants
Whitehouse Station, New Jersey

MARCEL DEKKER, INC. New York and Basel

Library of Congress Cataloging in Publication Data

Tedeschi, Robert J., [date]
Acetylene-based chemicals from coal and other natural resources.

(Chemical industries; v. 6)
Includes bibliographical references and index.
1. Organic compounds. 2. Acetylene. I. Title.
II. Series.
TP247.T43 661'.814 81-19607
ISBN 0-8247-1358-3 AACR2

MARCEL DEKKER, INC.
270 Madison Avenue, New York, New York 10016

Current printing (last digit):
10 9 8 7 6 5 4 3 2 1

PRINTED IN THE UNITED STATES OF AMERICA

PREFACE

This book examines the commercial applications, technology and newer developments related to acetylene, acetylenic compounds, and their derivatives. The author, during more than 25 years of involvement with acetylenic chemicals at both Air Reduction Company, Inc. (AIRCO)* and Air Products and Chemicals, Inc. (APCI), has observed that the triple bond combined with other functionality can confer unique properties upon a given compound or homologous series of compounds. This uniqueness often found diverse applications in the marketplace in areas such as fragrances, flavors, vitamins, drugs, surfactants, polymers, metal treatment formulations, pesticides, and adjuvants.

However, during the period 1950 to the present, the winds of technological change also inexorably foretold the decline of both calcium carbide and petrochemical acetylene as starting materials for multitonnage chemicals. Two patterns of change became apparent for acetylene and its derivatives: one adverse, related to the decline of acetylene as a source of commodity chemicals, the other positive and showing a significant growth of acetylenic chemicals and derivatives into varied application areas, some of large volume.

The significant economic and technological changes which affected acetylene directly from approximately the early 1940s to the present are detailed in this book. Commodity chemicals such as vinyl chloride, vinyl acetate, acrylonitrile, acrylates, chloroprene, and chlorinated solvents, which were manufactured over 30 years ago mainly from acetylene, are today made almost entirely from hydrocarbons such as ethylene, propylene, or butadiene. During this period, calcium carbide acetylene was displaced by petrochemical acetylene, and the latter gradually displaced by cheaper olefins or alkanes.

In Chap. 1, technology for the production of acetylene such as electric arc, regenerative furnace pyrolysis, combustion and plasma processes are discussed in terms of newer acetylene routes, raw materials, and competing technology. The possible renaissance of acetylene to some of its former prominence is also evaluated with respect to carbon-based acetylene processes. In the emerging age of coal in the United States this abundant raw material will be used both as an energy source and as a chemicals source. It is somewhat

*This company changed its name to AIRCO, Inc., and in 1978 was acquired by British Oxygen Corporation (BOC).

ironic that the earliest commercial technology for acetylene, the carbide process, was based on coal and limestone, both quite abundant raw materials in the United States. The calcium carbide route is still practiced in a few locations in the United States such as Louisville and Calvert City, Kentucky (AIRCO-BOC) and Astabula, Ohio (Union Carbide).

By the early 1960s the growth of Reppe chemicals anchored by an acetylenic base of butyn-1,4-diol and propargyl (1-propyn-3-ol) alcohol became significant. Reppe technology has its origins in the pioneering studies of Julius W. Reppe and his associates at the I.G. Farben in Germany. Novel chemistry and efficient pressure processes for a variety of acetylenic compounds and derivatives evolved from this important work. Today Reppe products such as butane-1,4-diol, tetrahydrofuran, N-methylpyrrolidone, N-vinyl-2-pyrrolidone and its polymers, vinyl ethers, and vinyl ether copolymers represent multimillion dollar markets and span a wide diversity of applications in the United States, Europe, and the Far East.

Concurrent with the growth of Reppe products has been the commercialization of secondary and tertiary acetylenic alcohols and diols via the base-catalyzed ethynylation of aldehydes and ketones. This diversified product line has exhibited superior growth in areas such as flavors, fragrances, drugs, vitamins, pesticides, pesticide formulations, surfactants, coatings, corrosion inhibitors, metal treatment aids, and polymers. These uses and the technology for the preparation of these acetylenics are discussed in Chap. 2.

I wish to acknowledge my debt to both Air Products and Chemicals Inc. (APCI) and the Air Reduction (AIRCO) Company for the opportunity of being exposed over the years to a branch of chemistry and technology that I have found to be both fascinating and personally rewarding. The management of APCI, in particular, has been both patient and skillful in developing this specialty line of non-Reppe acetylenics. For those who may ponder the "acetylenic connection" between both companies, AIRCO sold its chemical business and assets in 1971 to APCI, which included acetylenic chemicals as well as a line of polymer and plastic products. Shortly thereafter, the Acetylenic Chemicals Division was established as a separate business entity, which today is part of the new Performance Chemicals Profit Center of APCI.

I am grateful to the following APCI personnel who reviewed sections of the text and offered numerous suggestions: J. T. Barr, C. E. Blades, W. E. Daniels, H. P. Gallagher, C. A. Heiberger, H. L. Jaffe, G. Mantell, C. McKinley, T. J. Medovich, J. W. Pervier, and W. M. Smith.

Finally I wish to express special thanks to Loretta Kiersky, head of library and information services at AIRCO Central Research Laboratories, who prodded me to start this work and encouraged me to pass the point of no return. I dedicate this book to my children Marc, Lisa, and Thomas, to my wife Jean, and to those who would see in the triple bond a lever or springboard to the unique and the useful.

Robert J. Tedeschi

CONTENTS

1

ACETYLENE AND COMMODITY CHEMICALS

1.1 INTRODUCTION

Coal, and chemicals that can be made from it, will accelerate in growth against a world-wide scenario of ever-increasing inflation and oil shortages. From July to December 1979, the average price of a barrel of oil climbed from $15 to $28, a marked increase over the 1970 price of $1.80 [1]. A further price rise to the $32-36 level during 1981 is now fact, with single shipments of oil on the spot market already in the $40-45 range.

With the advent of "gas lines" in 1979 and a steady increase in gasoline prices to more than $1/gal, together with estimates as high as $2/gal by 1982-1983, it was apparent to Americans everywhere that gasoline and fuel oil, the offsprings of oil, were in tight supply, and that an oil crisis had ripened on a national and global scale. Price pressures from the OPEC nations and the crisis in Iran have accelerated worldwide inflation and fears of critical oil shortages in the years ahead. These factors will continue to reshape politics, and cause prices of oil-based chemicals to rise inexorably in the years ahead, in spite of temporary oversupply gluts in world markets.

Key articles [2,3] over the years have pointed out the possibility of serious depletion of liquid fossil fuel reserves throughout the world beyond the year 2000. Although natural gas reserves [4] are still projected as abundant, it is likely that the substitution of natural gas for oil in industrial uses will also put adverse pressure on this chemical-energy raw material.

It is generally conceded that coal, solar, and nuclear energy are the energy choices of the future. Coal is particularly important since it can also function as a starting material for the synthesis of organic molecules. Coal, used extensively in the early history of this nation and throughout the world, will again become important through the emerging technologies of coal gasification [4] and liquification [5] and by virtue of its abundant reserves.

There are obvious problems and challenges in converting an economy from its convenient and economic dependence on oil to the broad use of coal. However, the decision was made in July 1979 by President Carter to undertake a broad, federally subsidized program to free the United States from dependence on foreign oil. An important part of this program involved the use of coal for the production of liquid and gaseous fuels.

Unlike oil, there are no known processes to date for converting coal directly to commodity olefins such as ethylene, propylene, or butadiene, since coal gasification, liquification, or pyrolysis technologies yield only minor amounts of olefins [6]. Gasification produces over 90% carbon monoxide and hydrogen [6], which are converted by Fischer-Tropsch technology [7] to either methanol or hydrocarbons, depending on the catalyst used. The hydrocarbons or methanol must then be cracked by conventional techniques [8] to olefins. Coal liquification yields a mixture of tars and liquids which are mainly polycyclic in structure and also contain nitrogen and sulfur compounds. The complex mixture cannot readily be converted to simple olefins. Hence, commodity olefin production from coal is viewed as a multistep technology and is not competitive with direct cracking methods using simple hydrocarbons such as ethane, propane and butane, or more recently, crude oil [8].

In contrast, acetylene can readily be produced from coal by AVCO (coal + hydrogen plasma) or calcium carbide (coke + limestone) technologies. The latter process dates back to the origins of the acetylene industry and is still practiced by Air Reduction Company, Inc.-British Oxygen Corp. (AIRCO-BOC) at Louisville, Kentucky. The calcium carbide and AVCO processes are discussed in Sec. 1.2 and 1.3.2.

In the early 1960s the use of acetylene for the production of commodity chemicals (acrylonitrile, acrylates, chlorinated solvents, chloroprene, vinyl acetate, vinyl chloride) reached a peak of about 1 billion lb acetylene [9]. Since then, olefins such as ethylene, propylene, and butadiene derived from cheap cracking feedstocks have mainly replaced acetylene for these products. However, it is expected that future cracking plants (c. 1985) will utilize crude oil in preference to decreasing supplies of hydrocarbons, and that this will result in increased amounts of acetylene. The advanced cracking reactor (ACR) system [8] that Union Carbide Corporation is planning to put in production will coproduce 200 million lb of acetylene with 1 billion lb of ethylene from about 3.1 billion lb of feed. It may be preferable to use this large acetylene by-product source for chemicals manufacture rather than to hydrogenate it to ethylene, as is now done in the older cracking plants.

The U.S. production and supply of ethylene in the 1980s is characterized by much uncertainty and apprehension [10]. Typical problems that have been discussed are (1) nature and supply of feed stocks, (2) price, (3) merchant market supplies, (4) new capacity, (5) inventory supplies, and (6) reinvestment capital for new capacity. The investment cost for new ethylene capacity in the mid-1980s has been estimated at close to $1 billion, for which the return on investment is considered poor. The price of ethylene will rise from its 1979 price of 14¢/lb to 23 to 29¢/lb in 1984, depending upon the feed stocks and technology used.

Beyond the year 2000 it appears likely that oil, and perhaps natural gas, may be too expensive to use for chemicals manufacture. Under these conditions the use of acetylene could have a renaissance in the crucible of coal technology. The future of coal-based technologies is further amplified in Sec. 1.7.2.

In this book acetylene is discussed with respect to such topics as:

1. Processes for its production from natural gas, oil feedstocks, and coal
2. Acetylene properties, stability and handling
3. Guidelines for the safe use of acetylene
4. Production of commodity chemicals from acetylene and competitive processes
5. Production of large-volume Reppe acetylene-based products
6. Production of specialty acetylenic alcohols, diols, and derivatives

Reppe technology has its origins in the pioneering research of Julius W. Reppe and associates at I. G. Farben, in Germany. This novel chemistry and efficient pressure process is described in Chap. 2. Also discussed are other ethylation processes, such as liquid ammonia technology and the unique properties, applications, and markets for both large-volume and specialty acetylenes and derivatives.

1.2 CALCIUM CARBIDE ACETYLENE

The period 1920-1950 witnessed the rapid commercial growth of acetylene as a building block for large-volume chemicals [11], mainly monomers. The most economic and safe method of handling and shipping acetylene was and still is as calcium carbide, which is reacted with water in special generators to liberate acetylene (Ref. 11, Vol. 1, pp. 165-370 and Ref. 12). However, obvious hazards still remained in handling "carbide," and this soon resulted in commercial service sites where acetylene was manufactured and liberated from large acetylene generators and piped directly "over the fence" to nearby chemical plants. Large generator facilities at Calvert City and Louisville, Kentucky (Air Reduction Company); at Niagara Falls, New York (Union Carbide); and Schkopau, East Germany; were erected to supply the growing needs of such products as vinyl chloride, vinyl acetate, acrylonitrile, chloroprene, and chlorinated solvents both here and in Europe. The largest producers of acetylene via the calcium carbide route in the United States were Air Reduction Company (now AIRCO-BOC) and Union Carbide (Linde Division).

The carbide process [12,13] is operated at 2000°C. The electrical energy consumed is approximately twice the heat of reaction:

$$CaO + 3C \xrightarrow{\Delta} CaC_2 + CO \quad \Delta H_{298}, 111\ k_{cal}$$

Process economies have been introduced in recent years for the reuse of lime. Also, improved furnace design, which has helped reduce heat loss, makes possible the recovery of carbon monoxide and minimizes air pollution.

However, the large electrical power requirements of carbide acetylene dictating specific plant sites (cheap power), together with the less desirable aspects of handling large volumes of solids (coke, lime, calcium carbide), preordained its eventual decline. These factors in turn catalyzed research-development activities that led to new acetylene processes based on cheap, easily processed hydrocarbons available in large amounts from the rapidly growing oil industry. This industry, possessing large amounts of available capital for new ventures, decided during the early 1940s to enter the rapidly growing and profitable field of commodity chemicals.

During the next two decades huge petrochemical complexes were erected in the climatically and economically favored Gulf Coast area. Petrochemical acetylene based on the pyrolysis of low-cost petroleum fractions grew rapidly in importance at the expense of the static carbide industry, which was being phased out gradually due to its inability to compete effectively. By the period 1965-1975 most calcium carbide facilities had been closed down.

However, calcium carbide when viewed today and in the future in terms of its abundant raw materials, limestone and coal (coke), could again become an attractive technology as petroleum-based feedstocks become scarcer and more expensive, as discussed in Sec. 1.1. Also, the energy requirements to produce calcium carbide in the future may be supplied more economically by nuclear or fusion energy. The latter, available through the fusion of hydrogen isotopes to helium, is a potentially cheap and abundant energy source of the future for both this process and plasma technology, whereby coal is reacted with hydrogen to produce acetylene (see Sec. 1.3.2). The carbide process also can be regarded as primarily coal and water based since the calcium hydroxide resulting from acetylene formation (see Sec. 1.2.2) can be readily dehydrated to calcium oxide and recycled in the process. It is also likely that with improved furnace design and solids handling, further economics in this process are possible.

1.2.1 Technology Details: Acetylene Production from Calcium Carbide

Miller [11] has described in great detail the early history, manufacture, and important details concerned with the production of calcium carbide and of acetylene derived from it. Although the basic operations have not changed greatly in over 80 years, the technology is still relatively complicated, as it involves many processing steps and is characterized by much solids handling. The reader is referred to the excellent account by Miller.

1.2.2 CaC_2 Formation

The overall chemistry of the process, based on limestone and coke (a) + (b), is summarized below. The hydrolysis of calcium carbide (c) to acetylene is described further in Sec. 1.2.3

(a) $CaCO_3 \longrightarrow CaO + CO_2$ -41.8 kcal.

(b) $CaO + 3C \longrightarrow CaC_2 + CO$ -11.2 kcal

(c) $CaC_2 + 2\,H_2O \longrightarrow Ca(OH)_2 + C_2H_2$ +31.0 kcal

The endothermic reaction of quick lime (CaO) with coke proceeds in two stages in which calcium metal is an important intermediate.

(d) $CaO + C \longrightarrow Ca + CO$ -125.3 kcal

(e) $Ca + 2C \longrightarrow CaC_2$ +14.1 kcal

The overall reaction (d) + (e) is shown in (b).

The carbide reaction [11,19] requires the intense energy of the electric arc and is initiated somewhere between 1500-2000°C, depending upon the purity of the reactants and, the degree of contact. In recent years it has been customary to use compressed briquettes made of fine powder to increase productivity. The rate of calcium carbide formation is proportional to the area of contact between the quick lime and coke. However, the reaction is more complex than the simple equations indicate, and is believed to be characterized by solid solutions and intermediate compound formation in a changing phase matrix [11].

The molten carbide emerging from the furnace is not further purified, except for dust impurities blown off from the furnace by the evolved carbon monoxide or by ferrosilicon slag removed by an optional tap hole. The solidified carbide (ingots or granules) resembles carborundum, and in larger pieces does not constitute a serious hazard by being exposed to a reasonably dry atmosphere for short periods of time.

Technical carbide averages 79-83% CaC_2, 7-14% CaO, 0.4-3% C, and 0.6-3% SiO_2. Remaining impurities comprise Fe-Si, SiC, Fe_2O_3 CaS, $CuSO_4$, Ca_3P, Ca_3As_2 and aluminum compounds. An average acetylene gas yield from "carbide" is 300 liters gas kg solid carbide. Attempts have been made to correlate average material balance with reactants, products, and impurities in the carbide furnace. However, large variations have been noted, which are difficult to explain, and which are probably caused by variations in impurities. The average power consumption in more modern plants has been estimated at 3.0-3.5 kWh/kg (2600-3000 kcal/kg) CaC_2 produced.

The calcium carbide production facility of AIRCO-BOC is still in operation at Louisville. An important new application for CaC_2 is its use in steel production to lower the sulfur content in cheaper coke and avoid air pollution. The sulfur is converted to calcium sulfate instead of being evolved as toxic hydrogen sulfide. Calcium carbide is also shipped to AIRCO-BOC from the Louisville site to its Calvert City location, where it is converted to acetylene for the manufacture of acetylenic chemicals by General Aniline and Film Corporation (GAF) and Air Products and Chemicals, Inc., at nearby locations. At present the only production facilities in the United States for calcium carbide and acetylene derived from it are the AIRCO and former Union Carbide (Ashtabula, Ohio) facilities. In recent years air pollution from the carbide furnaces, visible as a dense brown smog, and caused by "fines," swept up by the evolved carbon monoxide, has created economic problems for producers in meeting air pollution regulations of the Environmental Protection

Agency (EPA). Carbon monoxide is also a toxic air pollutant which must be either recovered or oxidized (burned) to carbon dioxide. The latter approach is preferred at present.

1.2.3 Hydrolysis of Calcium Carbide to Acetylene

The reaction of calcium carbide with water is significantly exothermic.

$$CaC_2 + 2H_2O \longrightarrow Ca(OH)_2 + C_2H_2 \quad +31.0 \text{ kcal}$$

Pure calcium carbide on treatment with the stoichiometric amount of water at 18°C and 1 atm will evolve 484 kcal/kg carbide. An 82% commercial grade of CaC_2 (equivalent to 300 liters C_2H_2/kg.) will give a heat effect of 429 kcal, larger than expected due to heat of hydration of CaO in commercial carbide (~12% CaO).

The carbide-water reaction in the absence of cooling will heat the reaction mass to over 700°C. At such elevated temperatures acetylene will undergo exothermic reactions, further increasing the hazards of the process. Without suitable cooling an acetylene vapor explosion will eventually result, particularly if air is present. The carbide mass has been observed to become incandescent if uncontrolled by water cooling.

In commercial practice the heat of reaction is controlled by the use of various acetylene generators and techniques.

1.2.4 Carbide-to-Water Generator

Granular carbide is fed continuously into excess water, and the slurry is agitated by gas evolution and a mechanical agitator. Generators of this type operate at about 60-80°C and pressures of 15-30 psi. Water is used in a 6:1 excess over carbide, and spent slurry is continuously discharged as fresh carbide is added and acetylene removed.

In the United States, the medium-pressure (30 psig) generator system is preferred over the low-pressure (15 psig) system. Generators are generally designed [15] so that they have an acetylene generating rate of 0.03 m^3 (1 ft^3)/hr using 0.5 kg carbide and 3.8 liters (1 gal) water. Although there are numerous generator designs, all are characterized by common features such as a stirred water-containing reactor chamber, carbide hopper, carbide feed (screw or hopper) device, acetylene water scrubber, and gas storage tank. The generators are protected with flash arrestors to prevent acetylene deflagrations or vapor explosions and have pressure-relief valves to keep the pressure below 15 or 30 psi. Automatic controls (mechanical, pneumatic, electrical) also are used to monitor and control the safe operation of the process. Acetylene storage tanks are maintained below 15 psig.

Carbide feed granules of the proper size are important in preventing clogging or jamming of valves and in aiding the sinking and thorough mixing of carbide in the reaction slurry. The preferred carbide size is 14 ND (6 × 2 mm). If the particle size is too small (reactive carbide) or the reaction chamber does not have a stirrer (not customary), then oiled carbide is used to control the rate of reaction and maintain safety.

Large generators have acetylene capacities of 90,000 ft^3/hr, and can process about 150 t carbide before cleaning out the system is required. This "wet process" generation of acetylene is, or was, at least, popular in the United States.

1.2.5 Dry (Lime) Generators

This system comprises a cylindrical generator containing a series of interconnected trays to which the carbide is fed continuously to the top tray. The charge gradually moves to the bottom of the generator as it is sprayed with a limited amount of water and mixed (radial blades) between trays. The well-mixed carbide (almost dry) drops to the bottom of the generator and is collected in a boot prior to steaming and recovery. Acetylene is removed from an exit line located near the bottom tray. About 1 lb water/lb carbide is used in this method.

The exothermic carbide-water reaction is controlled at about 93°C by using a limited amount of water, which is converted to steam, together with efficient mixing in each tray section to prevent local over heating. These type of generators have an advantage over the carbide-to-water types in that lime can be recovered dry for reuse or sale. They have been extensively used in Europe in chemical manufacturing.

1.2.6 Water-To-Carbide Generators

The original miners', bicycle, and carriage lamps utilized the simple principle of dripping water slowly on to the carbide mass and burning the evolved gas. This type of generator has been extensively used for small portable generators, but is less important for large commercial installations.

Miller [11,14] has described numerous generator systems with both reactor diagrams and photographs. Since the early 1900s many articles and patents have documented different systems for generating acetylene from carbide. Up to possibly the early 1960s this technology (both carbide and acetylene production) was practiced worldwide.

1.2.7 Lime and Coke Starting Materials [11]

In general, carbide production facilities have prepared calcium oxide (quick lime) by decomposing ("burning") a 98% grade calcium carbonate ($CaCO_3$) to CaO and CO_2. To prepare a satisfactory commercial grade of carbide, both the purity of the $CaCO_3$ and the CaO must be high (99% conversion to CaO). Both horizontal rotary kilns and vertical shaft kilns are generally used. Magnesium oxide impurity in the CaO is particularly objectionable since it increases both carbon and electrical power usage in the process.

The carbon most commonly used in Europe for carbide manufacture is metallurgical coke of 86-89% carbon, screened to a particle size of 6-25 mm. In the United States, at one time, petroleum coke was mainly used due to its higher purity. However, it has proven to be less reactive and more expensive over the years, and is seldom employed.

In past years (1940-1960) the hydrated lime [$Ca(OH)_2$] resulting from large-scale carbide-to-water acetylene generators was dumped in large lime ponds or impoundments. Since lime recycling was not practical then and sales through miscellaneous uses were slow, huge lime deposits accumulated at carbide acetylene sites. Today, at the few carbide sites left in the United States, the lime is both recycled and sold, so that the problem has lessened. Increased legislation (federal, state, and local) which also restricts the disposal of carbide lime hydrate has increased sales. Typical uses for this lime include waste-water treatment, neutralization of spent pickling acids, soil neutralizer, and sand lime bricks.

1.2.8 Process Hazards

The operation of the carbide furnace, with the exception of electrical and high-temperature hazards, is not considered a dangerous operation. Calcium carbide can be stored readily and safely in covered steel containers, and at one time was also shipped extensively in closed hoppers or drums.

The principal hazard associated with carbide is its accidental exposure to water, high-moisture air, or steam. The high heat generated in the carbide mass, if it is partly or totally enclosed, will ultimately cause acetylene to explode or detonate. Many earlier explosions (early 1900s) attributed to undiluted liquid acetylene were probably initiated by the hydrolysis and ignition of carbide.

1.3 PETROCHEMICAL ACETYLENE

A large amount of work describing new acetylene processes has been patented during the last 30 years. The recent books of Miller [11] (British Oxygen Company) are particularly noteworthy in documenting acetylene technology, properties, and commodity chemicals derived from it.

Below are described the principle acetylene processes using hydrocarbon feedstocks. In recent years, excellent review articles [13,14] and an ever-increasing volume of patent literature (Ref. 11, Vol. 1, pp. 371-475) provide convenient sources of information. The processes can be classified into six main categories, summarized in Table 1-1 together with coal-based and by-product sources of acetylene.

1. Electric arc (Badische Aniline Soda Fabrik [BASF]
2. Regenerative furnace pyrolysis (Wulff)
3. One-stage combustion (BASF, Sachsse)
4. Two-stage combustion (Union Carbide, Montecatini)
5. Plasma arc or jet (Chemische Werke Huels AG)
6. New processes

A common principle of technology characterizes all petrochemical processes. A hydrocarbon feedstock is subjected to an intense energy source whereby it is heated to 1200-1500 K, using short residence times (0.01-0.001 sec). The cracked gas, containing acetylene, is quickly cooled (milliseconds) to about 550 K. Acetylene in the 1500 K range has a lower free energy (greater stability) than other hydrocarbons and its formation is favored. Rapid water quenching of the hot gases prevents acetylene and the uncracked feedstock from decomposing into carbon and hydrogen.

The high energy requirement to initiate the decomposition of hydrocarbons is shown with methane:

$$2CH_4 + 174,000 \text{ Btu} \longrightarrow C_2H_2 + 3H_2$$

The optimum temperature for cracking methane is 1500 K. However, with increasing chain length of the hydrocarbon, lower temperatures are possible.

Table 1-1 Acetylene from Natural Gas, Petroleum, and Coal Sources

Typical Process	Principal Feedstock	Technology	Typical Companies
Electric arc	Methane, gas oils	Arc or plasma	Huels, Du Pont
Sachsse	Methane, natural gas mixtures	Partial combustion (one stage)	BASF, Dow, Monsanto
SBA	Methane, natural gas mixtures (1st stage); naphthas, heavier feedstocks (2nd stage)	Partial combustion (one and two stage)	SBA, M. W. Kellogg
Wulff	Natural gas, naphthas heavier feedstocks	Regenerative furnace pyrolysis (four cycles)	Union Carbide, Wulff
Montecatini	Natural gas, naphthas	Partial combustion under pressure	Montecatini, Diamond Alkali
By-product acetylene	Ethane, hydrocarbons, naphthas, oil	Steam cracking	Major oil and chemical companies (EXXON, Shell, Dow, Union Carbide)
Calcium carbide	Limestone ($CaCO_3$) and Coke	CaC_2 from C + $CaCO_3$; C_2H_2 from CaC_2 + H_2O	AIRCO-BOC, Union Carbide
AVCO	Coal and hydrogen	Hydrogen plasma	AVCO, pilot and demonstration plants)

1.3.1 Processes

Electric Arc Process   This process was first pioneered by BASF in 1920, and a commercial plant was in operation in Germany (Huels) in 1940. In the United States, manufacture by this method started in the early 1950s.

An advantage of the arc process was that the energy could be applied very rapidly to the reactant, and the conversion of hydrocarbons to acetylene was significantly higher than with regenerate or partial combustion processes. However, the arc process was also known to be sensitive to process variations, which could lead to large amounts of undesirable carbon formation when operating poorly. E. I. du Pont de Nemours and Company, Inc. operated a modified arc process at Montague, Michigan prior to 1971 to provide acetylene for chloroprene (neoprene rubber) manufacture. The Du Pont facility, designed to provide cheaper acetylene than was available from the carbide process, was closed down by 1973 due to poor process economies.

Regenerative Furnace Pyrolysis (Wulff) Process   The Wulff process (Ref. 11, Vol. 1, pp. 384-391 and Ref. 13, pp. 181-193) has a high degree of flexibility since by controlling the feedstock and operating conditions, the process can be made to deliver mainly either acetylene or ethylene. The regenerative furnace functions by heat exchange of the cracked gases with the refractory material of the furnace. Hydrocarbons are generally cracked at 1150°C and 0.50 atm using steam as diluent. After acetylene has been separated from the product gases, part of the residual gas is burned to reheat the refractory alumina furnace. The process involves four cycles of about 1 min each, which in turn comprise pyrolysis and heating in one direction of flow, followed by a reversal of direction. The process works best with propane or light naphtha feedstocks.

Deposition of tars and carbon is not a problem in the Wulff system since they are burned off as fuel. Also, the regenerative nature of the process enables most of the sensible heat of the cracked gas to be recovered (heat exchange with the refractory oven) minimizing acetylene decomposition. A Wulff facility has been in operation since 1952 at Los Angeles, California, with a rated capacity of about 1 million lb acetylene. Wulff technology has been also practiced in Europe, South America, and the Far East.

Combustion (One- and Two-Stage) Processes   Both the one- and two-stage processes [13] involve the combustion of the feedstock to yield the necessary energy to attain the required 1200-1500°C range. The one-stage method (BASF or Sachsse process) involves the partial combustion of preheated (500°C) methane with oxygen via a specially designed distributor and burner. Variations of this process have been patented [11,13].

The two-stage method burns process off-gases or lower value fuel gases in the first stage. Air or oxygen is used in the first stage of combustion, and exit gases reach a temperature of about 2000°C before entering the second-stage reaction zone. Here, the hydrocarbon feedstock is mixed with the entering hot gases and is pyrolized to acetylene and ethylene. The hot exit gases in both processes are stabilized by rapid water quenching.

The partial-combustion technique is as popular as the Wulff method and was used by Union Carbide, BASF, Montecatini, Rohm and Haas, Dow Chemical U.S.A., and American Cyanamid Corporation. Combined yields of acetylene and ethylene can reach 50%, and the ethylene/acetylene ratio can be varied from 0.1 to 3.

Plasma Methods The so-called plasma arc methods (Ref. 11, Vol. 1, pp. 405-406) have been extensively studied at the pilot plant level, but at present do not represent important commercial routes. In one variation [16], hydrogen preheated to about 1000°C is decomposed in an electric arc to generate hydrogen atoms. The hydrocarbon feed is then injected into the plasma of H radicals outside the arc, where as they are further cooled, they recombine to molecular hydrogen and release the energy which cracks the hydrocarbon to acetylene. By this method, propane can be converted in 80% yield to acetylene.

In a related technique, the plasma jet [17], argon is employed as the plasma source. Its use is claimed to give very high yields at residence times of about 1/2 msec, but the power requirements are also quite high. At present, hydrogen is believed to be the most attractive source of plasma gas, since on cooling from 5000K to about 1200K, it liberates about 135 Kcal/mol, which is sufficient to form 1 mol of acetylene from methane. Although, Huels has been operating a plasma arc process in West Germany since 1940, this technique at present is considered too costly in terms of power requirements for general use.

1.3.2 Technology Details: Acetylene from Hydrocarbons

The decomposition (cracking) of hydrocarbons at elevated temperatures (>1000°C), using various energy sources and processes, has been the subject matter of many studies, publications, and patents, spanning roughly a period from the early 1920s to the

1960s. It is only possible in this book to briefly describe this important worldwide technology, in terms of the most important processes. The excellent treatise by Miller [11] provides details of numerous processes for thermally converting hydrocarbons to acetylene, together with voluminous references covering both publications and patents. A review published in 1978 by Kirk and Othmer [18] also provides a recent update on the technology of acetylene production from hydrocarbons.

Acetylene at 1500 K is more stable than other alkanes or alkenes (methane, ethane, ethylene, etc.). This is a consequence of the well-known fact that the free energy of acetylene decreases at elevated temperatures compared to other hydrocarbons. However, this advantage is very short-lived (milliseconds) in the environment of a high-temperature (1500-1700 K) pyrolysis furnace or electric arc reactor. Hence, the process "trick" to stabilize acetylene is to rapidly cool (water quench) the emerging reactor gases below 500-550 K to prevent decomposition to carbon, hydrogen, or acetylene condensation products (tars). Also, to convert an alkane into acetylene the feedstock must be subjected to temperatures in excess of 1200°C for very short periods, then rapidly quenched.

An interesting calculation of theoretical energy requirements for converting hydrocarbons to acetylene was reported by Miller (Ref. 11, Vol. 1, p. 393), based on the following general equation, and a reaction temperature in which K_p for the reaction was 100 (Table 1-2):

$$\text{(a)}\quad \frac{2}{n} C_nH_{2n+2} \rightleftharpoons C_2H_2 + (1 + \frac{2}{n} H_2) : K_p\ 100$$

These theoretical energy values do not include the energy required to preheat the feedstock so that the exothermic reaction will proceed. Preheating by natural gas and liquid fuels is considered cheaper than complete dependence upon electrical energy in the case of an arc reactor.

The data show that with increasing chain length the energy requirements are less to convert hydrocarbons to acetylene. Also, typical data for heats of formation (kJ/C atom at 1500 K) of various hydrocarbons divided by the number of carbon atoms in the hydrocarbon, support the above data and conclusions: methane (-92.4), ethane (-53.9), n-octane (-31.6), carbon or hydrogen (0), ethylene (18.1), acetylene (110.1).

It is likely in future years that increased amounts of higher molecular weight feedstocks will be used in cracking processes to

Table 1-2 Conversion of Hydrocarbons to Acetylene, Theoretical Energy Values

Hydrocarbon	ΔH° (25°C)[a] kcal	ΔF°[b] T	°C[c]	H_1-H_{298}[d] kcal	H.R.[e] kcal/mol
Methane	90.0	95.4-0.0640	1470	54.4	144.4
Ethane	74.4	78.2-0.0638	1160	34.5	108.9
Propane	70.7	73.7-0.0358	1090	29.6	100.3
n-Hexane	67.5	69.7-0.0635	1010	25.0	92.5
n-Decane	66.1	68.0-0.0635	980	23.4	89.5
Ethylene	41.7	43.9-0.0316	1680	38.8	80.5

[a]Heat of reaction for Eq. (a).
[b]Free energy change at TK.
[c]Reaction temperature at which acetylene is formed.
[d]Sensible heat.
[e]Heat required per unit of C_2H_2.
Source: Ref. 11, Vol. 1, p. 393.

produce ethylene and propylene. This will in turn result in increased amounts of by-product acetylene, which may have to be utilized for commodity chemicals manufacture (see Sec. 1.1 and 1.3.2).

However, whether electric arc technology or other thermal processes will be used to produce acetylene and olefins, such as ethylene and propylene, will depend upon overall process economics, particularly energy costs. The arc process in general requires about 11 kWh energy to produce 1 kg of acetylene, while the older carbide process averages 9.5-10.5 kWh.

Electric Arc Technology Although electric discharge processes can supply high cracking temperatures rapidly, with resultant higher acetylene yields compared with other processes, they have not been a favored technology in the United States. Processes such as regenerative furnace pyrolysis (Wulff) or one- and two-stage combustion methods have been preferred during the period 1940-1970. The high cost of electrical energy, the poor recovery, and the use of excess heat, together with large amounts of by-product carbon formation may account for the unpopularity of this method in the United States. However, based on extensive technological expertise [19] the I. G. Farbenindustrie at Huels, Germany put a plant into operation (15 arc reactors) in 1940 rated at over 200 t/day acetylene, which was still in operation in the 1970s. The plant [20] was modified for 17 arc reactors, and is now claimed to have a capacity of 100,000 t/yr acetylene.

The arc is struck between a graphite cathode and an annular copper anode. The temperature of the arc is 20,000 K, which provides an average temperature in the reactor of about 1700-1800 K ($\sim$1500°C). The hydrocarbon feed, generally natural gas mixtures or methane, is subjected to the intense energy of the arc for 1-2 msec before being rapidly cooled (quenched) with water below 500 K. The energy requirements to produce a kilogram of acetylene average about 11 kWh/kg of acetylene produced.

Du Pont has developed a modified (rotating) electric arc reactor which is schematically shown in Figure 1-1. This reactor was used in their Montague, Michigan plant [21] (capacity 50 million lb/yr) to produce acetylene for the manufacture of chloroprene and neoprene rubber. The plant came on stream in 1963 but recurring problems with excess carbon formation resulted in the facility being closed down in 1972, in favor of a more efficient butadiene-based process for chloroprene (see Sec. 1.6.7).

The preferred feed for the reactor is methane [22], although gas oils can also be used. The cathode is a carbon rod, and the anode

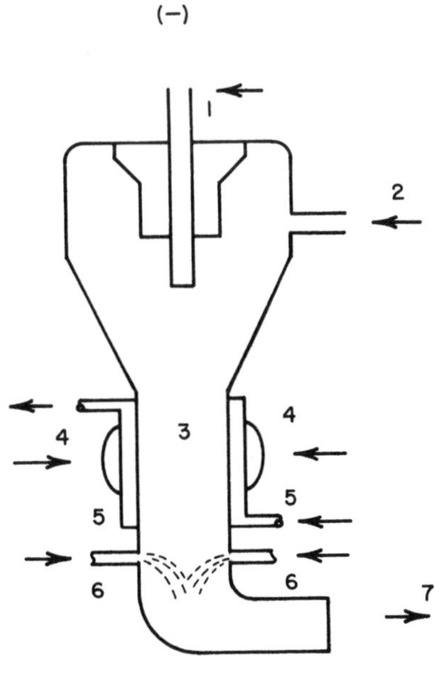

1. Cathode feed
2. Feed gas
3. Anode tube
4. Magnet coil
5. Anode coolant
6. Water quench
7. Cracked, quenched effluent

Figure 1-1 Du Pont electric arc reactor.

a concentric copper tube which is water cooled. The arc burns from the end of the cathode into the anode tube, and the magnetic field causes the arc to rotate at 8000 r/min in a trumpet-like configuration. A reactor cracking methane at the rate of 54 kg/hr. operates at 335 V (dc) and 1000 A. The process is claimed to yield an acetylene concentration of 18% in the cracked gas. The rotating arc is claimed to eliminate uneven corrosion of the cathode, which is advanced automatically to compensate for electrode wear. The water-cooled anode was also claimed to lessen carbon deposition at the arc root. The average power consumption of the system was 10.6 kWh/kg acetylene. A modified process [23] involved heating methane in the rotating arc to 1600°C, then quenching the cracked gas with propane to 1100°C, followed by a rapid quench to 300°C.

The Huels electric arc furnace [11,13] is shown in Figure 1-2. The arc (G) is struck between the cathode (C) and the grounded anode pipe (H). It averages about 100 cm in length and extends

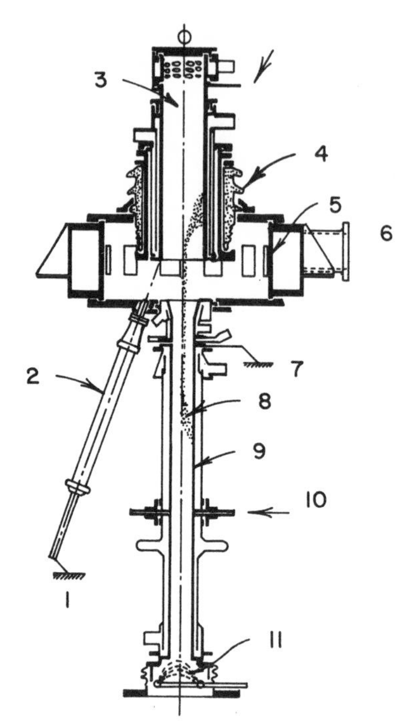

1. Ground
2. Ignition electrode
3. Bell shaped cathode
4. Insulator
5. Turbulence chamber
6. Gas tube cracked
7. Ground
8. Arc
9. Anode pipe
10. Liquid hydrocarbon injection point
11. Quench water

Figure 1-2 Huels electric arc furnace.

about 50 cm into the anode pipe (1.5 mx 100 mm). The gas feed, consisting of fresh hydrocarbon and recycle gas, enters the turbulence chamber (H) where it is given a rotary motion before entering the anode pipe. The cathodic or anodic starting point of the arc moves freely upward or downward, and in the upper section of the anode pipe the arc takes the rotary motion of the feed gas.

The hot ($\sim$1750 K) reactor gases are prequenched with cold hydrocarbon (C_2 or higher) about 20 cm below the arc (anode), which results in partial cracking of the hydrocarbons, mainly to ethylene. In this manner, depending upon the degree of prequenching, the ratio of acetylene to ethylene can be controlled. The emerging gas effluent is cooled with a water spray below the anode to about 450 K.

All parts of the reactor with the exception of the insulator (E) are made of iron. The electrodes are water cooled and have a wall thickness of 10 to 20 mm. The arc is operated at 8000 kW and 7000 V using a direct current of 1150 A.

The important components of the feedstocks (including recycle gas) and reactor effluent for the arc process are shown below.

	Volume %	
Hydrocarbon	Feed Gas	Effluent Gas
Methane	53	17
Acetylene	1	16
Ethylene	2	7
Ethane	10	1
Propane	8	1
Propylene	2	1
Butane	13	2
Hydrogen	3	50
CO	1	1
N_2	3	1

Partial Combustion Processes This technology has been practiced extensively in the United States. It is also known by such designations as flame, one-stage, and two-stage partial combustion processes. The basic principles involved are discussed in Sec. 1.3.1. A preferred feedstock has been methane or C_1-C_3 hydrocarbon mixtures, although naphtha can also be used. The more important processes utilizing the partial combustion principle are:

1. BASF (Sachsse)
2. Société Belge de l'Azote et des Products Chimiques de Marly (SBA)
3. Montecatini

A number of reactor furnace modifications and processes such as the Hoechst HTP submerged arc, MAFKI, Grienko, Nardac, Dow Chemical U.S.A., Eastman Kodak, and Phillips have also been described by Miller [11], as well as numerous process patents covering the last 40 years on a worldwide basis.

The partial combustion method generally involves preheating a natural gas or methane feedstock with a limited amount of oxygen (hydrocarbon/oxygen ratio ~ 2) to about 500-650°C. Burner design is of critical importance to prevent preignition, backfiring, flame blow-off, and excess carbon deposition. Combustion must be rapid and uniform throughout the entire reactor chamber, followed by rapid quenching of the reactor gases. Oxygen (98%) is preferred over air since it not only gives a more rapid reaction rate, but also provides more concentrated acetylene in the cracked gas. Also, the by-product hydrogen and carbon monoxide left in the off-gas after separation of acetylene can be used as reactants to produce methanol or ammonia, thereby improving overall process economics.

The one-stage combustion process is typified by the Sachsse or BASF method [24]. About two-thirds of the hydrocarbon is burned to provide the thermal energy needed to crack the remaining feed to acetylene. In the two-stage process the combustion and pyrolysis (reaction) chambers are extensions of each other. In the combustion chamber an off-gas or hydrocarbon of low fuel value is burned in oxygen with superheated steam. The combustion results in a short ring of flame surrounded by a steam-containing atmosphere. The feed to be cracked (often naphtha or heavier hydrocarbons) is injected just below the flame into the hot combustion gases, where it is cracked to acetylene, then rapidly water quenched. The second-stage process can utilize a wide variety of feeds (C_1-C_3 feeds, naphthas, and hydrocarbons up to C_{15}) and is more versatile than the one-stage route, which is limited to gaseous hydrocarbons. The SBA process utilizes both one-stage and two-stage methods, and is shown in Fig. 1-4. The two-stage method resembles regenerative furnace pyrolysis (Wulff) technology, since it can produce both acetylene and ethylene in wide ratios without modifying the reactor system.

BASF (Sachsse) process I. G. Farben operated one of the first Sachsse [24] acetylene production facilities (1942) at Oppau, Germany. A typical cracking facility could produce about 1.2 t/hr of acetylene at a concentration in the cracked gas of 8 to 9% C_2H_2. In 1942, there were five such units at Oppau. BASF, formerly part of the I. G. Farbenindustrie, carried on this technology [25] after the dissolution of the I. G. at the end of World War II. In the United States this technology became known as the BASF-Chemico process.

The feed to an acetylene burner comprised 4000 m^3/hr methane and 2250 m^3/hr oxygen. The reactants were preheated to 450-500°C with flue gas, using counterflow-tube and shell heat-exchangers. Although methane has been preferred, the BASF process can also use naphtha feedstocks or (LPG). Figure 1-3 shows a typical

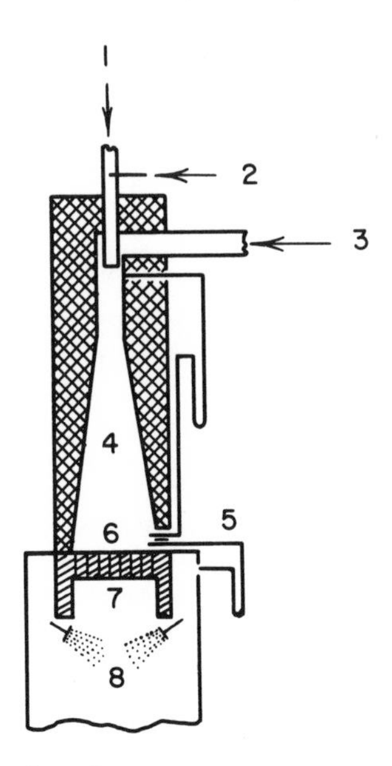

1. Oxygen
2. Inert gas (safety purge)
3. Methane or naphtha feed
4. Neck and mixing chamber
5. Auxiliary oxygen
6. Burner block
7. Reaction chamber
8. Water quench

Figure 1-3 BASF (Sachsse) burner.

burner design for the BASF process. Under normal conditions about one-third of the methane feed is cracked to acetylene, while the remainder is burned. Preheating methane to 650-700°C will cause preignition and is to be avoided. With naphtha the preheating temperature is about 320°C.

The oxygen-hydrocarbon reactants are preheated and introduced separately. They are premixed in a venturi chamber (Fig. 1-3) before passing to the multichannel burning block. The gas velocity through the burner channels is kept high enough to prevent backfiring but low enough to prevent extinguishing of the flame. The stability of the flame is increased by a small booster stream of auxiliary oxygen at the burner block. The contact time in the reactor chamber is several milliseconds, followed by immediate water quenching below 500°C.

The cracked gas, prior to removal of acetylene, varies in composition with respect to either methane or naphtha.

Cracked Gas	Methane		Naphtha
Acetylene	8	98.2%	9
Hydrogen	57		43
Carbon monoxide	26		37
Carbon dioxide	3		4
Methane	4		5
Ethylene	0.2		0.2
Other (N_2, O_2, hydrocarbons)	1.2		1.2

SBA (one- and two-stage) processes The Société Belge de l'Azote et des Produits Chemiques de Marly (SBA) erected a plant in 1953 at Marly, Belgium for the production of acetylene from methane (one-stage process) [26]. Since then a number of plants have been built around the world, which include modified (two-stage) technology in which liquid hydrocarbons are added below a flame to produce both acetylene and ethylene. In Sec. 1.3.1 and 1.3.2 the underlying principles of the SBA technology are discussed. M. W. Kellogg-SBA has been the agent for this technology in the United States and the British Commonwealth nations. In the United States it is commonly known as the *SBA-Kellogg process*.

The SBA burners used for one-stage and two-stage combustion processes are shown in Fig. 1-4. The type I burner (one-stage) is similar in its operation and results to the BASF system. The type II burner (two-stage), although resembling the type I burner, is modified to accommodate a combustion and reaction chamber as well as a feed system for liquid hydrocarbons (naphtha). The preheated naphtha is mixed with superheated steam and introduced to the reaction chamber just below the flame. The ratio of ethylene to acetylene can be varied from 0.2 to 2.0. Typical compositions of cracked gases for the one- and two-stage processes are

One-stage (type I burner) C_2H_2, 8%; H_2, 57%; CO, 25%; CH_4, 5%; N_2, 0.5%; CO_2, 4%; $C_3 + C_4$ hydrocarbons, 0.3%

Two-stage (type II burner) C_2H_2, 10%; C_2H_4, 4%; H_2, 43%; CO, 18%; N_2, 4%; CO_2, 11%; higher paraffins, 0.2%; other acetylenes, 0.7%; olefins, 0.5%; aromatics, 0.1%

The preheated feed for one-stage reactors is primarily, methane or a C_1-C_2 mix with oxygen. Two-stage burners use either fuel gas (coke-oven effluent) or the cracked gases from either reactor, once the acetylene has been removed. The feed to be cracked to acetylene is usually naphtha, but this can be varied widely up to C_{10}-C_{15} hydrocarbons. Both oxygen and the hydrocarbon feed are heated separately to about 650°C before introduction to the burners. Preignition can be a problem over 600°C, but with the use of superheated steam diluent for the hydrocarbon (two-stage process) this is less of a problem. A recommended residence time in the reactors is 2.3 to 2.9 msec at 1400-1500°C. A complete water cooling and quenching system for both burners is described by Miller [11] for the combustion, reaction, and gas quench zones (Fig. 1-4).

Montecatini process The Montecatini process was developed in conjunction with natural gas deposits discovered in northern Italy [27]. In 1955 a plant was erected at Novara with a capacity of 20 t/day. The reactor system is similar to that of BASF (Sachsse), except that the process is operated under a pressure of 4 to 6 atm.

Although it is well known [11] that acetylene decomposition is accelerated under pressure at high (1400-1700°C) temperatures, it was observed that at higher flow velocities, the acetylene yield was comparable to results at atmospheric pressure. Also, the use of 2 vol % steam in the feed eliminated the danger of preignition. Some advantages claimed for this technology are:

1. Compression costs are less, since natural gas and acetylene-ethylene cracked gases are handled under pressure for purification and chemicals manufacture.
2. Burner size is much smaller than atmospheric or lower pressure units, yielding increased productivity.
3. Energy savings are possible in preheating the burner and in heat exchange equipment due to the higher heat exchange (coefficients) capacity of the compressed gas. Energy required for pyrolysis is highly concentrated, and thermal losses are minimized.
4. More efficient use of the heat content of the system, particularly for the recovery of acetylene in later steps via absorption in cold methanol (-20 C), followed by boiling methanol to remove acetylene.
5. Wetting and removal of carbon black from the burner by the use of superheated (>100°C) water (lower temperature water is ineffective).

The important burner block in the reactor is composed of high-alloy steel channels embedded in a ceramic arc [27,28]. The

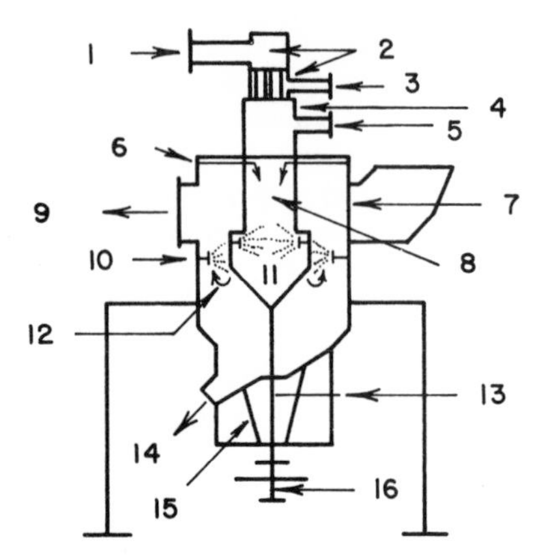

1. Methane
2. Mixer
3. Oxygen
4. Grate
5. Oxygen
6. Screen water
7. Safety bursting disc
8. Combustion chamber
9. Pyrolysis gas
10. Cooling water
11. Quenching
12. Pyrolysis gas
13. Quench water
14. Water outlet
15. Support
16. Water quench control level

Figure 1-4 SBA partial combustion burners used for one-stage and for two-stage combustion processes.

remainder of the unit resembles a Sachsse burner as shown in Fig. 1-5. The average residence time in the reactor is several milliseconds at 1500°C. The hot gases can be quenched, first with naphtha or light mineral oil, followed by water. The use of naphtha facilitates additional formation of both acetylene and ethylene, making this system similar to the SBA two-stage process (cf. Fig. 1-4). The compositions of cracked gases with and without the naphtha prequench are:

Light Naphtha Quench: C_2H_2, 19%; C_2H_4, 30%; H_2, 58%; CH_4, 20%; CO, 10%; CO_2, 3%; C_3H_6, 1%; (total, 140% of water quench).

Water Quench Only: C_2H_2, 9%; H_2, 54%; CO, 25%; CH_4, 7%; CO_2, 4%; C_2H_4 + hydrocarbons, 1.7% (total 100%).

In the present-day economic environment, in which acetylene production is tied to other hydrocarbons such as ethylene and

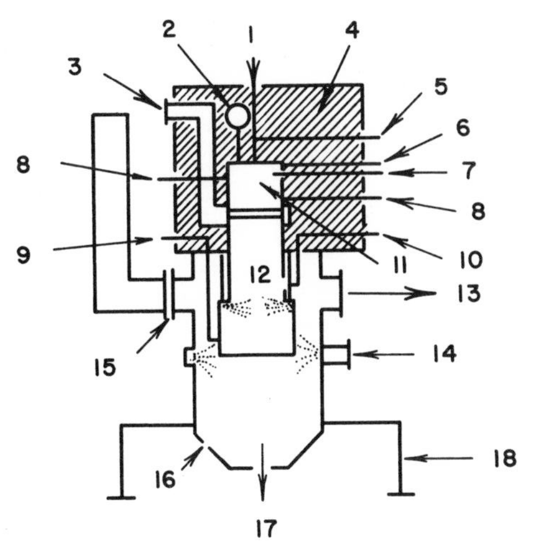

1. Steam
2. Oxygen
3. Naphtha
4. Heat circulation
5. Combustion gas
6. Steam
7. Ignition port
8. Cooling water
9. Quenching water
10. Steam
11. Combustion chamber
12. Reaction chamber
13. Pyrolysis gas
14. Quenching water
15. Safety bursting disc
16. Peep hole
17. Used water outlet
18. Support

Figure 1-4 continued. SBA partial combustion burners used for two-stage combustion processes.

propylene, the use of naphtha and higher molecular weight feeds via a two-stage or prequench operation may be preferable to partial combustion or electric arc technology. The Wulff process (regenerative furnace pyrolysis), which can deliver wide ratios of acetylene and ethylene (see Sec. 1.3.2) also is attractive technology.

Regenerative Pyrolysis (Wulff) Process This technology [29,30] is probably second in importance to partial combustion processes and has been practiced both in the United States and in Europe. Some advantages claimed for this process are:

1. Feedstocks can be varied from methane, propane, and other natural gas mixtures to naphthas and, finally, heavier hydrocarbon oils with good results.

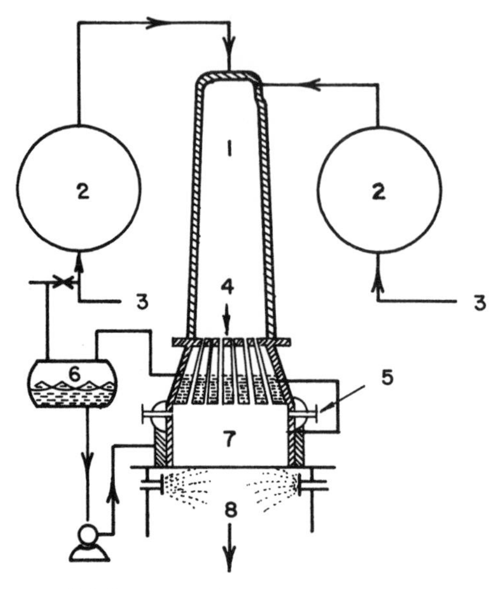

1. Mixing chamber
2. Preheater
3. Natural gas feed
4. Burner block
5. Igniter
6. Cooling water
7. Reaction chamber
8. Water quench and effluent

Figure 1-5 Montecatini burner.

2. The acetylene/ethylene ratio can be varied widely, with a 1:1 ratio attractive for chemicals manufacture.
3. The regenerative nature of the process allows recovery of most of the sensible heat in the cracked gas.
4. Air is used in place of oxygen, and tars and carbon are burned off in the process and do not accumulate in the furnace.

The regenerative furnace is shown in Fig. 1-6, and is made of high-purity refractory tile containing cylindrical (checkers) channels for gas flow. The important elements of the furnace are right and left plenum chambers, refractory checkers, fuel gas burners, and a central combustion space in which fuel is periodically injected to provide energy for heating the system. The process operates on a four-cycle basis (summarized below) in which the directions of flow of the cycles are reversed.

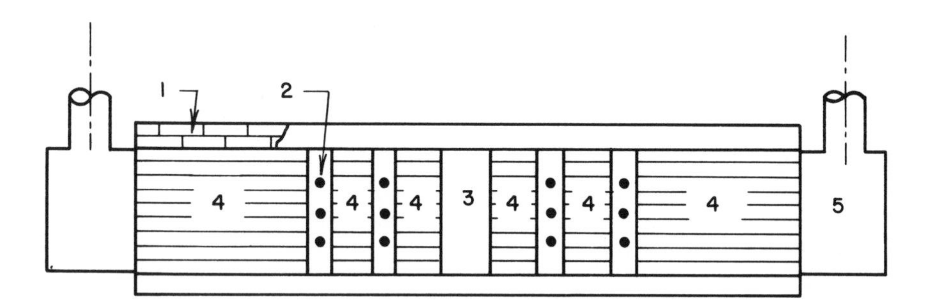

1. Double layers of fire brick
2. Gas burners
3. Combustion chamber
4. Refractory checkers
5. Plenum chamber
6. Cross section view-refractory checkers

Figure 1-6 Wulff (regenerative pyrolysis) furnace.

Cycle 1 Air enters the right plenum and is heated to 1100°C. (from previous cycle). At the fuel injection point, air is mixed with the fuel (off-gas free of acetylene and ethylene, or part of the cracked gas). The heat of reaction raises the temperature of the furnace to 1400-1500°C and burns off previously deposited tars and carbon as the reactor gases emerge from the left plenum at 450°C.

Cycle 2 The feedstock, diluted with steam and off-gas at 0.5 atm, is fed into the furnace from the left plenum and reaches cracking temperature (1400-1500°C) at the furnace midsection and is then cooled to 450°C by the refractory tile before emerging from the right plenum. The residence time is 0.1 sec, of which 0.03 sec is at the temperature maximum. Some carbon and tar is deposited on the right side of the furnace, and the emerging gas is further cooled by a water spray and "cleaned up" by a Cottrell precipitator before acetylene is recovered.

Cycle 3 Air fed to the left plenum is heated by the checkers and, when mixed with fuel, burns off deposited carbon and tars from the right side of the furnace.

Cycle 4 The feedstock enters the right plenum and cycle (2) is repeated, but in an opposite direction of flow.

The entire four-cycle sequence requires 1 min. Although methane requires a temperature of 1500°C for cracking, higher hydrocarbons can be cracked at 1200°C. The composition (vol %) of a cracked gas from propane is C_2H_2, 10%; C_2H_4, 4%; CH_4, 15.0%; H_2, 56%; CO_2, 7%; N_2, 5%; CO, 2%; O_2, 0.7%.

Small amounts of benzene (0.4%), propylene (0.2%), allene (0.2%), vinyl acetylene (0.2%), ethane (0.1%), diacetylene (0.04%), propane, and alkynes (trace) compose 1.1% of the total cracked gas. Total yields to acetylene and ethylene average 51-59% at C_2H_2/C_2H_4 ratios of 4:1 and 1:4, respectively.

Acetylene from Olefin Production It is known that in the cracking of hydrocarbons to ethylene and propylene, a potentially large by-product source of acetylene is present. Yields of acetylene from naphtha cracking can average 0.2-0.6% depending on the severity of the cracking. With gas-oil feedstocks the acetylene content can be as high as 1%. It has been estimated (Ref. 18, pp. 231-232) that if total world production of ethylene was 35 Mt/yr., based on an average 30 wt % ethylene content in cracked crudes, then an acetylene content of 0.5 wt % would amount to 580 kt/yr. The present world production of acetylene is estimated at about 800 Kt/yr.

However, there are problems in being able to utilize this acetylene source. The fact that this by-product source is scattered over many industrial sites in different countries makes large-volume use of this potential acetylene source difficult. Also, most olefin producers have continuous fixed-bed hydrogenation units in their production facilities where the more active acetylene in the feedstock can be selectively hydrogenated to mainly ethylene.

Although this potential by-product source of acetylene is of interest to companies producing specialty acetylene chemicals, the companies most likely to utilize this source would be the large olefin producers themselves. This situation could take place if coal-based acetylene (see the following section) began to compete with ethylene and propylene for the manufacture of the commodity chemicals discussed in Sec. 1.6 and 1.7. That this event will occur in the next 5 to 10 years is unlikely.

It is also likely that by the mid-1980s the amount of by-product acetylene available from cracking will increase significantly. A recent review [8] of cracking technology for the production of ethylene projects shows that new cracking facilities operable by 1985 will utilize crude oil in preference to lower hydrocarbon feeds. This will yield much greater amounts of acetylene. Both Union Carbide and

Dow are operating multi-million pound demonstration plants. The ACR system [8] to be utilized by Union Carbide will yield the following co-product distribution (MM lb) via the production of 1 billion lb of ethylene from about 3.1 billion lb of crude oil: acetylene (200), propylene (135), butadiene (90), BTX (benzene, toluene, xylenes) (435), pitch (583), fuel (600), other (7). This large source of acetylene, added to existing by-product acetylene production from present-day crackers, could again initiate the manufacture of commodity chemicals from acetylene. Other interesting by-product acetylenes from this source are methyl and ethyl acetylenes, which have interesting specialty uses in fragrances and flavors, drugs, and polymers.

Acetylene from Coal and Hydrogen The reaction of finely divided coal or carbon with hydrogen in a hydrogen plasma is described in Sec. 1.3.1.4. Although research on this route has been and still is being conducted worldwide [31,32], the AVCO [33] process in the United States is the technology most likely to be scaled up to commercial production. The work has been sponsored mainly by the Office of Coal Research (OCR).

Figure 1-7 is a schematic of the hydrogen plasma arc reactor developed by AVCO. The arc rotates and is spread out radially in the reactor due to the external magnetic field, and somewhat resembles the Du Pont arc reactor (Fig. 1-1). The finely divided coal crosses the arc area, is activated, and reacts with the hydrogen plasma at a temperature gradient of 8,000 to 15,000 K. Acetylene yields as high as 33 wt % and acetylene concentrations up to 16% have been reported. Energy requirements as low as 9.5 kWh/kg acetylene have been claimed. The process consumes about 67% of the coal, and the remainder is recoverable as a char that could possibly be used for its fuel value.

Although the AVCO process has been claimed [34,36] to be superior to the partial oxidation of methane, there are varied opinions that, unless cheap energy (e.g., nuclear) is available, coal-based acetylene may be unable to compete with ethylene derived from the less energy-intensive cracking of hydrocarbons for commodity chemicals production. The economics of coal-based acetylene will also be tied to world reserves of both natural gas and liquid fossil fuels. Although oil reserves have been estimated to become depleted beyond the year 2000, new discoveries may push this date substantially into the future. The large oil potential of both Alaksa and Mexico is also just being felt. However, it is certain that once oil reserves do become significantly depleted and, hence, more expensive than coal, the latter will surely be used for the production of acetylene and the commodity chemicals derived from it.

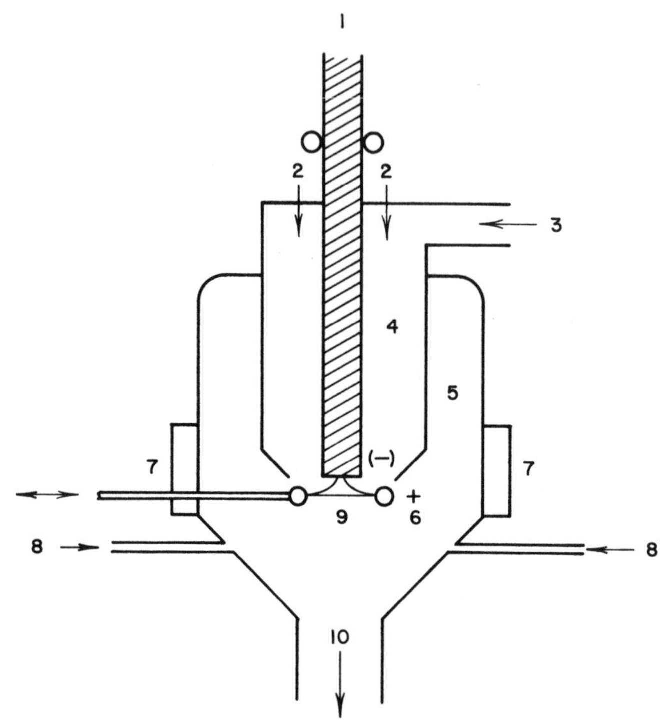

1. Cathode and drive rolls
2. Hydrogen feed
3. Coal feed
4. Plenum
5. Reactor body
6. Anode
7. Field coils
8. Hydrogen quench ports
9. Rotating arc
10. Char and gas to cyclone

Figure 1-7 Coal-hydrogen (AVCO) plasma-arc reactor.

During 1980 AVCO announced [35] an important breakthrough involving the coal-based arc process for acetylene. A 1-MW coal-fed reactor capable of producing 2 million lb/yr of acetylene was successfully demonstrated at Wilmington, Massachusetts via a joint project involving AVCO and the Department of Energy (DOE). Test results showed an acetylene yield above 35% based on coal with low electrical energy usage.

AVCO claims the pilot unit can be scaled to commercial size to provide acetylene competitive with, or cheaper than, ethylene for the production of commodity monomers such as vinyl chloride and vinyl acetate. Initial mid-1980 manufacturing cost comparisons between the coal-acetylene and ethylene routes for a 725 million lb/yr vinyl chloride (VCM) complex are summarized below based on cited transfer prices of acetylene and ethylene.

	Costs (MM dollars)				
Raw Material	Plant Inv.	Capital	Net Oper.	Transfer Price (¢/lb)	Pricing of By-products
Acetylene	68	105	98	17.1	At chemical value
Acetylene		126		26.1	At fuel value
Ethylene	81	132	140	25.0	At fuel and chemical value

VCM Transfer Prices at 12% Return Rate

Raw Material	By-product Pricing at	Transfer Price (¢/lb)
Acetylene	Chemical value	16.0
Acetylene	Fuel value	20.0
Ethylene	Chemical and fuel value	22.5

If the AVCO-DOE pilot studies can be successfully scaled up, and if petroleum-based feedstocks continue to escalate in price, it is likely that both AVCO and calcium carbide technology will become important before 1990.

A "wild card" in the oil-economics game is the intense political and inflationary price pressure being placed on oil and its derived products by the Arab and Third World nations. These factors in turn could expedite the introduction of coal on the world scene as a chemical and acetylene raw-material source. The future of coal as a feedstock for acetylene production and the use of fusion energy to realize this goal are discussed in Sec. 1.1, 1.2, and 1.7.2.

1.3.3 New Processes

Research [36] in plasma technology has shown that coal injected into a hydrogen plasma can be converted to acetylene in yields of

30 to 35% with a total energy consumption of 4 to 5 kWh. It has been estimated that this method could produce acetylene for about 5¢/lb prior to 1970. This new technology, with further improvement, has the potential of again making acetylene a preferred feedstock for such commodity products as vinyl chloride, vinyl acetate, acrylonitrile, and chloroprene.

Cheap coal (strip mining) and hydrogen (synthesis gas) could be attractive longer range feedstocks for this new plasma technology if current yields and energy usage are improved. Furthermore, long-range estimates of coal versus liquid fossil fuel reserves favor coal (cf. Sec. 1.1, 1.7.2). Beyond the year 2000, it is likely that the rapid depletion of oil reserves in the United States will adversely effect the use of oil-based hydrocarbons for both chemical and energy needs. Coal, present in much greater reserves, is already being used in gasification technology by natural gas companies.

Careful kinetic studies on the pyrolysis of methane by Happel and Kramer [37] have shown that it is possible to obtain high yields of acetylene where the main by-product is hydrogen, instead of tars and coke. Their studies also showed that pyrolysis in the presence of hydrogen suppressed carbon formation. During 1965, economic studies [38] indicated that acetylene could be produced at 3.5 to 5.0 ¢/lb at a level of 50 million lb/yr. Although the process was claimed to be operable at the pilot plant level, it has not yet been reduced to commercial practice. It is believed that Arab countries are evaluating this technology (Seabord Annual Petrochemical Conference, Abu Dhabi, 1976) for the conversion of natural gas (presently being flared) to acetylene, for the manufacture of vinyl chloride, acrylates, vinyl acetate, and acetaldehyde [39]. The combination of abundant, low-cost natural gas and cheap labor might make the classical acetylene-based routes for commodity chemicals attractive to Arab nations and Iran, particularly for internal use.

An improved "wet-wall reactor" for the cracking of hydrocarbons to acetylene and ethylene has been developed by the Royal University of Technology (KTH) in Stockholm, Sweden [40]. The reactor utilizes a hydrogen burner as the energy source, and it is possible to collect the product gases without dilution by combustion gases. The hydrogen produced by the cracking reaction is sufficient to supply the requirements of the hydrogen burner. Acetylene and ethylene yields of 25% and 54%, respectively, have been realized. The process, however, requires scale-up studies to check and possibly improve its performance.

Although nuclear energy has received strong citizen condemnation and an unfavorable public press after the Three Mile Island nuclear accident, it is still likely that this energy source will be expanded in the future. The use of both nuclear (fission) and fusion energy as heat sources for plasma and calcium carbide technology is discussed in Sec. 1.2, 1.3.2., and 1.7.2.

1.4 ACETYLENE PURIFICATION

One advantage of the carbide method is that it delivers high-purity (99.6%) acetylene [11] via hydrolysis of CaC_2. The relatively small amounts of impurities can be removed by solid absorbents (molecular sieves, alumina, etc.). In contrast, acetylene derived from pyrolysis processes is contaminated with CO_2, CO, N_2, H_2, HCN, H_2S, unreacted hydrocarbons and higher acetylenes, and the acetylene content is generally below 25%. When higher hydrocarbons are used, the cracked gas stream also contains olefins (ethylene, propylene, isobutylene), dienes, (propadiene, butadiene, isoprene), aromatics (naphthalene, benzene), and higher hydrocarbons, further complicating the separation problem. The acetylene purification train is a complicated, expensive, multistage unit which, however, when operated efficiently can ultimately deliver acetylene comparable in purity to the carbide process.

The acidic nature of acetylene and its ability to hydrogen bond [41] or complex [42] with Lewis base-type solvents (NH_3, DMF, THF, acetone, etc.) provides the key means of effecting separation. Some solvents that are used commercially to separate acetylene are water (Huels), liquid ammonia (SBA), cold methanol (Montecatini), N-methyl pyrrolidone (BASF), and DMF (Du Pont). Water and methanol are poor acetylene solvents, but can be made more effective at lower temperatures and under pressure. Water will form a solid acetylene-water complex [42] under pressure below 25°C.

Liquid ammonia, if used at low temperatures (-35° to -50°C) or in a pressure system [42,43] at ambient temperatures, is capable of dissolving large amounts of acetylene. It should be possible to use this concentrated (30-50%) acetylene-ammonia solution for a variety of chemical operations [44]. The much lower solubility of alkanes, olefins, hydrogen, carbon monoxide, etc., in liquid ammonia makes this technique feasible. The resulting synthesis-grade acetylene should be cheaper than welding-grade gas as a chemical raw material. Tenneco, Inc. utilizes liquid ammonia to recover acetylene from its hydrocarbon (Cat Cracking) operations at Houston, Texas.

1.4.1 Technology Details: Acetylene Processing

The recovery, purification, and drying of acetylene from calcium carbide is a relatively simple operation compared with the processing of the petrochemical product. In the latter case acetylene must be recovered from a process stream in which it occurs at levels of only 8 to 15% and where it is one of at least 10 other components. The purification train for a "petro" acetylene plant is relatively complex and considerably larger than the reactor sections used to produce acetylene.

1.4.2 Calcium Carbide Acetylene

Although commercial calcium carbide is relatively impure (∿80% CaC_2), acetylene liberated from "carbide" by water treatment is 99 to 99.8% C_2H_2 on a dry basis (Ref. 11, Vol. 1, pp. 331-342). Most of the gaseous impurity is often air with minor amounts of other hydrides such as phosphine (PH_3), arsine (AsH_3), and hydrogen sulfide (H_2S). A typical analysis of impurities in dry carbide acetylene is shown in Table 1-3.

The impurity content in acetylene is dependent upon the type of calcium carbide used and the temperature and method of generation.

1.4.3 Acetylene Purification

Impurities such as PH_3, AsH_3 and H_2S can be oxidized readily by the use of ferric chloride deposited on kieselguhr. Small amounts of $HgCl_2$, MnO_2, or $CuCl_2$ are added as cocatalysts. The ferric chloride composition can be periodically reactivated with air. This method has been used in carbide plants since the origins of the industry (Ref. 11, Vol. 1). Commercial ferric chloride compositions are sold under various trade names such as Regenetal and Catalysol.

Concentrated sulfuric acid can be used to absorb hydrides such as PH_3 and water, but the absorber should be cooled to avoid overheating and the possible ignition of SO_2 formed.

Washing acetylene with 80 to 100% sulfuric acid (scrubbing towers) has been common practice in carbide acetylene and chemical plants. Often, prior drying through a silica gel column is combined with sulfuric acid scrubbing.

Table 1-3 Average Acetylene Impurities (ppm-vol %)

NH_3	PH_3	AsH_3	H_2S	VA[a]	DVA[b]	DVS[c]	R-SH[d]
20	400	6	250	400	Trace[e]	700	Trace[e]

[a]Vinyl acetylene
[b]Divinyl acetylene
[c]Divinyl sulfide
[d]Mercaptans (R-SH)
[e]Acetaldehyde is sometimes present in traces

Another widely employed method is the use of chlorine water or hypochlorites as oxidants for hydride and sulfur impurities. This method is used as part of a countercurrent washing train that uses dilute (4-20%) H_2SO_4, chlorine water (0.1-0.2%, and caustic soda (7-10%) in the order listed. The acetylene is given a final purification over activated carbon.

Final drying of acetylene is important both for cylinder filling and when the acetylene is used to manufacture chemicals. By using a combination of refrigeration (0 to -20°C) and adsorbents such as alumina, silica gel, or molecular sieves, 99.99% of the water content in acetylene can be removed, with a final acetylene purity of at least 99.9%.

1.4.4 Recovery and Purification of Petrochemical Acetylene

Since acetylene in a cracked-reactor gas stream contains on average 10 to 15% acetylene, compared with 98 to 99% crude acetylene from the CaC_2 process, the purification system is considerably more complex. A common principle characterizes all petro-purification systems, namely the use of polar solvents to dissolve acetylene by hydrogen bonding [41]. Some of the more commonly used solvents are dimethyl formamide (DMF), N-methyl-1-2-pyrrolidone, butyrolactone, and liquid ammonia. Poor acetylene solvents such as methanol and water can be made fairly effective at lower temperatures and/or under pressure. Pressure and lower temperatures make highly polar (aprotic) solvents even more effective.

The following are various petrochemical acetylene processes and the solvents which are used to concentrate acetylene:

Process	Solvent	Temperature (°C)	Pressure (atm)
Electric Arc (Huels)	H_2O		18-19
Wulff	DMF		2
Montecatini	methanol	-70	Atmospheric
SBA	Anhydrous NH_3	-70	1
BASF	N-methyl-2-pyrrolidone		10

In Sec. 1.3.2 the numerous components of cracked acetylene streams are discussed. If not for the acidic nature of acetylene and its ability to dissolve (complex) in polar solvents, it is doubtful that an economical and safe acetylene process from hydrocarbons would have been developed. In spite of this solubility advantage the average cracked stream contains carbon black, tars, higher acetylenes, benzene, naphthalene, higher hydrocarbons, ethylene, propylene, butadiene, propadiene, carbon monoxide, carbon dioxide, nitrogen, and a high hydrogen (∿45-55%) content. With the exception of higher acetylenes and possibly carbon dioxide, most of the other components have low solubilities in the polar solvents (Lewis bases).

The separation and utilization (ethylene, propylene, CO, and H_2) of part of this mixture, together with the upgrading of acetylene purity to the 98-99% level, results in a rather complex separation technology (Ref. 18, pp. 227-231). This is further complicated by the "black panther" nature of acetylene, which under conditions of higher pressure (>15 psig) and higher temperature (>100°C) and conditions of static charge generation (gas flow in larger diameter pipes) can detonate or explode (cf. Sec. 1.4.7). Below are described several typical acetylene separation systems.

1.4.5 Wulff Acetylene Recovery System

In Sec. 1.3.2 the regenerative nature of the Wulff pyrolysis furnace is described. The quenched and cleaned (Cottrell precipitator) cracked-off gas is processed further [45,46]. In the separation

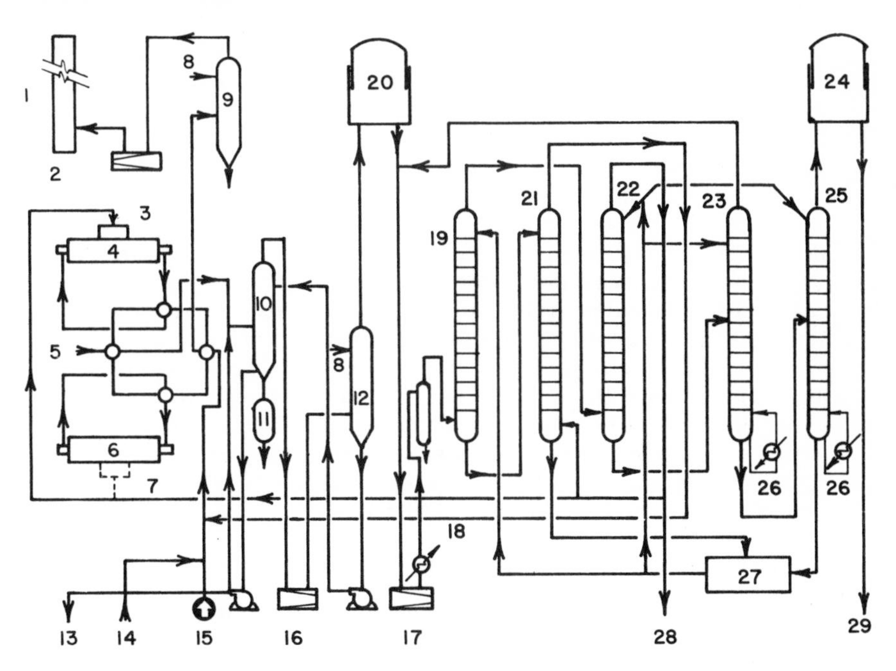

1. Stack
2. Vacuum pump
3. Wulff furnaces
4. Reheating cycle
5. Air
6. Cracking cycle
7. Fuel gas
8. Water
9. Combustion gas cooler
10. Lower quench
11. Tar collector
12. Cracked gas cooler
13. Quench to spray pond
14. Dilution stream
15. Fresh feed to furnaces
16. Vacuum pump
17. Compressor
18. Recycle loop
19. Diacetylene by-product absorber
20. Cracked gas holder
21. Diacetylene stripper
22. Acetylene absorber
23. Solvent stabilizer
24. Product (acetylene) gas holder
25. Lower boiling products
26. STM recycle loop
27. Solvent purification
28. Off gas
29. Acetylene product

Figure 1-8 Wulff process and acetylene recovery system.

system shown in Fig. 1-8 the process gas is compressed and then extracted with a limited amount of DMF solvent to extract out diacetylene and higher acetylenes. The hazardous diacetylene (decomposes in air) is stripped from the solvent with residual process gas and burned for its fuel value.

The gas stream from the diacetylene absorbers is extracted with the main DMF stream which forms a DMF-C_2H_2 solution. This solution is heated (stripped) to remove less-soluble gases, which are compressed and recycled to the main absorber. The undissolved residual gas is water scrubbed and then recycled for its fuel value. The DMF-C_2H_2 solution is heated (70-100°C) to free acetylene, which is sent to storage at 12 to 15 psig. The higher acetylenes are removed from the DMF by heating and stripping with additional residual gas. Polymers and tars solubilized in the DMF are kept to a low value by taking a small solvent slip stream and diluting with water to precipitate by-products. Both DMF and water-DMF mixtures are purified by distillation.

1.4.6 SBA Acetylene Recovery System

The SBA or SBA-Kellogg process utilizes the highly polar solvent liquid ammonia [41,11,46] as the extractant medium for acetylene. The quenched cracked gas is made free of CO_2 by washing with dilute aqueous ammonia and sodium hydroxide solution. Aromatic impurities are then removed by extraction with kerosene. Although liquid ammonia under pressure at ambient room temperature has a high solubility for acetylene, and in the proper acetylene:ammonia mole ratio can be handled safely, it is generally used at -70°C and at atmospheric pressure to extract out acetylene. Acetylene is desorbed from the C_2H_2-NH_3 complex by heating to 25-40°C under pressure and washing the gas free of ammonia with water and dilute acid. At lower temperatures, a small amount of dissolved ethylene is separated before desorbing the acetylene. Higher acetylenes and other by-products can be separated as process bottoms from the distilled ammonia. Tenneco utilizes liquid ammonia in its petrochemical acetylene plant at Houston, Texas.

1.4.7 Acetylene Stability and Handling

A painting [47] by the artist A. Kent depicts acetylene as a black panther, before a background of flames. The analogy with the feared cat is well taken, since one can work or use acetylene for years in assumed safety, and then an explosion may suddenly result, generally through human error or equipment failure. The energy-rich nature of acetylene can be readily appreciated when one reflects that the detonation of a pilot-sized reactor can set up shock waves which are detected 2 to 5 miles from the source. Today, the technology of purifying, storing, transporting through special pipelines, and using acetylene (Ref. 11, Vol. 1, pp. 484-542) in chemicals manufacture is well understood and explosions and detonations are rare.

The Germans during World War II observed that acetylene could not be transported safely at even low (1-2 atm) pressures in wide-diameter (50-300 mm) pipes without incurring the risk of explosion [48]. When small-diameter (<50 mm) pipes were packed with steel or porcelain Raschig rings or bundles of small-diameter steel tubes the danger of explosion and detonation was largely eliminated. Acetylene could be transported by such techniques across distances of over 80 km safely. The use of acetylene to make butyn-1,4-diol, an intermediate for butadiene and rubber production in Germany during the war years, is described in Sec. 2.1.1 to 2.1.3.

From the thermodynamic viewpoint, acetylene is unstable with respect to carbon and hydrogen, the elements of its birth. When decomposition is triggered by an energy source (heat, electrical or static charge, compression and pressure), a huge amount of energy (53,500 cal/g at 18°C) is liberated. Under adiabatic conditions a temperature rise of close to 3100°C and a pressure rise of about twelvefold is possible. If the decomposition is very rapid and a shock wave is sustained, as in a long tube (>30 m), then a detonation results which magnifies the pressure by up to 200 times the calculated pressure.

Much physical data [11] has been accumulated over the years concerning conditions which will decompose, explode or detonate acetylene. Most of these test explosions have been induced by electrically fusing various wires in pure acetylene or in acetylene mixtures at various pressures, and in vessels or pipes of different diameters and lengths. These tests have given rise to such terms as *initiation*, *limiting decomposition pressure* and *temperature*, *deflagration*, *detonation*, and *predetonation distance*, to measure and define the characteristics of acetylene decomposition. The tests have been well documented by Miller [11], Copenhaver and Bigelow [48] and by Kirk and Othmer [18].

In spite of many studies on the stability of acetylene, "the black panther" is still somewhat of an enigma, particularly with respect to early 1900 industrial explosions. Before international legislation restricted the use of pure acetylene to 15 psig and its transport to cylinders filled with acetone and a porous solid filler (kieselguhr and balsa wood), acetylene was commonly transported as liquefied gas or solid (-80°C) cakes. It was also used as a source of illumination on railroad passenger cars. There are no documented cases of such cylinders blowing up railway cars. Many of the early industrial acetylene explosions have been attributed to the triggering effect of water-calcium carbide fires. The transport and storage of calcium carbide is also carefully regulated by law, particularly with respect to the container. Miller (Ref. 11, Vol. 1, pp. 484-542) has described in detail the numerous regulations regarding handling, use, and transport of acetylene.

1.4.8 Factors Leading to Acetylene Decomposition and Detonation

Initiation   To initiate acetylene decomposition, an external energy source and sometimes, catalytic agents, must be present. Typical initiators are heat, pressure, electric spark, static charge, fusion of metal wires, copper, silver, iron rust and metal scale. Copper and silver can form the corresponding acetylide derivatives, particularly under ammoniacal conditions, and these materials, being unstable, can initiate the decomposition of acetylene. Copper and silver tubing, fittings, adapters and welds must be avoided in handling acetylene. The best metals for use with acetylene are stainless or rust-free carbon steel vessels, tubing, and fittings.

Air introduced into an acetylene environment with a combination of the above ignition sources will also cause explosion. Acetylene in the presence of copper and silver catalysts (acetylides, oxides, halides) at elevated pressures and temperatures can be polymerized to polyacetylenes (cuprenes), vinyl acetylenes, diacetylene and higher acetylenes. In the presence of air, these materials, particularly diacetylene and cuprene, are hazardous. Polyacetylenes particularly, if contaminated with copper acetylide will ignite or smolder in air, while diacetylene will exothermically decompose with much carbon formation (black smoke). Chloroacetylene if suspected should be accounted for, because even in low amounts it is highly hazardous and will explode in air. Chlorine and acetylene can react explosively in the vapor state.

Deflagration   The first stage of an initial acetylene decomposition or explosion is a deflagration, in which the initial flame travels (10-100 meters/second) through the unburned acetylene at less than the speed of sound. The pressure rise is gradual and continuous, and generally averages 8-14 times the initial pressure. Deflagrations can generally be contained, quenched, or controlled by the use of narrow reactor vessels, tubing, flame arrestors, and dilution of acetylene with inert gases (N_2, CO_2, CH_4).

Detonation A detonation results when a deflagration is unchecked, particularly where long process pipe of wide diameter (>50 mm) is used. The resulting flame or wave front increases in velocity as fresh acetylene is consumed until it becomes supersonic in velocity ($\sim$1800-2000 m/sec) and a detonation results [11,18]. Concurrent with a large energy release (about 53,000 cal/g), the pressure rises 200-fold over the initial value. The detonation is characterized by a discontinuous jump in pressure at the wave front, and the force of the explosion is highly directional. It is possible to blow a relatively small hole through a pressure vessel without affecting the bolts and other accessory equipment. Also, the shock wave may be heard for miles in one direction, while nearby the sound of the detonation may not be detectable. Once conditions leading to detonation have been underway for some time, it is generally not possible to contain the detonation in larger pilot or plant equipment.

Predetonation Distance The *predetonation distance* is defined by the length and diameter of the pipe used, together with the initial acetylene pressure. It is also a measure of the conditions which lead from a deflagration into a detonation, with the resultant large (100- to 200-fold) increase in pressure observed. It is the distance the flame travels before becoming a detonation. Table 1-4 summarizes typical results of ignition studies that determined predetonation distances for various pipe lengths and internal diameters with respect to the initial pressure which can lead to a detonation (Ref. 11, Vol. 1, pp. 484-542).

These data show that pure acetylene cannot be transported safely in long transmission lines at pressures in excess of 3.5 atm without increasing the risk of explosion.

Maintaining Safety in Acetylene Transmission Pipes Extensive tests by the Germans at Griesheim before and during World War II showed that acetylene could be safely transported long distances if internal pipe diameter was less than 25 mm (1 in.) for pressures of several atmospheres [48]. If nitrogen were used as an inert

Table 1-4 Predetonation Distances

	Lengths of 2.5 cm pipe			
Initial pressure (atm)	3.5	3.8	5.0	20.0
Predetonation length (m)	9.1	6.7	3.7	1.0

Varying pipe diameter (cm)	Pipe length (m)				
4		40	20	20	
8	60	40	30	20	
16					120
	Initial pressure (atm)				
	8.3	9.0	13	16	21

diluent at the 50% level, the safe pressure limit could be raised to 3 atm.

For the more practical use of larger diameter (50-450 mm) pipes, it was found that by packing the larger pipe with bundles of smaller diameter (10-13 mm) pipe, detonations could be completely prevented up to 3 atm. By the combination of packed tubes, inert diluents (N_2, CO_2, CH_4) at the 50% level, and the possible use of acetylene solvents, acetylene, over the years, has been handled safely in chemical processes both in Europe and in the United States.

Pressure-Temperature Effects Exposing acetylene to high temperatures in the absence of an ignition source can lead to decomposition and explosion. At one atmosphere pressure, decomposition begins at 635°C, while at 2 atm the decomposition temperature averages 510-540°C. The minimum temperature (hot spot) for acetylene decomposition is lowered by increasing the pressure.

Below are shown decomposition pressure-temperature studies [11] in which acetylene was decomposed by instantaneous fusion of a platinum wire (0.5 mm diameter):

°C	Decomposition Pressure (atm)
15	1.40-1.60
50	1.40-1.48
100	1.25-1.33
140	1.10
150	1.13
180	1.06

Numerous decomposition studies have also been carried out by initiating decomposition by the use of fused wires having different melting points. It is apparent that the intense local heat of the melted wire is effective in decomposing acetylene at relatively low pressures

Metal	Pt	Fe	Cu	Al
Melting point (°C)	1760	1530	1080	660
Decomposition pressure (atm)	1.4	1.7	1.8	2.7

Inert Gas Diluents The use of a wide variety of gases, inert to acetylene, in the absence of catalysts, was studied as part of the German war effort to handle large amounts of acetylene safely [48]. Typical of the gas diluents studied were hydrogen, carbon monoxide, carbon dioxide, nitrogen, methane, ethylene, and propane. For the stabilization of acetylene in the 5- to 20-atm range, propane and methane were, respectively, 2 and 1.5 times as effective as nitrogen in preventing decomposition. Today, however, nitrogen is probably the most commonly used diluent.

Ammonia [43] also has been claimed to be a good stabilizer for acetylene and is used as the extractant media for acetylene in the SBA-Kellogg process. Excellent stabilization has been observed with carbon monoxide [49], which is of great importance since C_2H_2-CO mixtures are reactants for the commercial production of acrylates (see Sec. 1.6.2). A 20:80 C_2H_2-CO mixture, with a C_2H_2 partial pressure of 46 atm at 19°C, could not be detonated by the fusion of platinum wires. Nitrogen used at the 50% level prevents the decomposition of acetylene at initial C_2H_2 pressures of 6 atm. Carbon dioxide gives similar stabilization at a 42% concentration.

Figure 1-9 shows the relationship of the limiting decomposition pressure of acetylene mixtures plotted against the acetylene content in the mixture at 15°C. A comparison between nitrogen and methane as stabilizers at the 50% level shows the superiority of methane (∿225 psi) over nitrogen (∿140 psi). Propane would be expected to stabilize at ∿280 psi on the graph shown in Figure 1-9.

Also, the limiting decomposition pressure of acetylene-nitrogen mixtures versus % C_2H_2 in the mixture is shown in Figure 1-10 at temperatures of 15, 100, and 200°C. A comparison of 30% and 50% C_2H_2-N_2 mixtures is of interest in showing the increased stability noted at 15°C by comparing the 15 and 200°C curves.

1.4.9 Liquid Acetylene

Liquid acetylene is no longer of commercial importance, since it cannot be legally transported in the undiluted form at pressures of 30 to 50 atm. Although it can be detonated by the usual fused-wire techniques, it has been regarded as no more dangerous than common explosives such as TNT or picric acid. Tedeschi and coworkers successfully used liquefied, undiluted acetylene [42] as both reactant and solvent up to its critical temperature (35°C) in such reactions as ethynylation [50], metalation [51], carbonation [51], and complex formation with transition metal salts [42]. The reactions were carried out in a barricaded steel cubicle, using small (25-50 ml) amounts of liquid acetylene in special reactors [42]. During a research program spanning almost a year, no decompositions or explosions of liquid acetylene were noted. Although this work showed that liquid acetylene could be handled far more safely than a hazardous, explosive material such as liquid ozone, it is still not a recommendation to use undiluted liquid acetylene commercially. The enigma of acetylene's black panther nature will probably never be adequately explained.

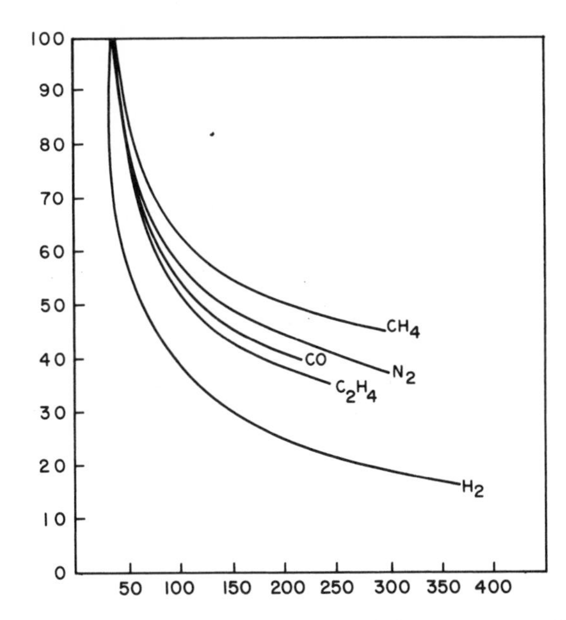

Fig. 1-9 Limiting decomposition pressure for acetylene-diluent mixtures.

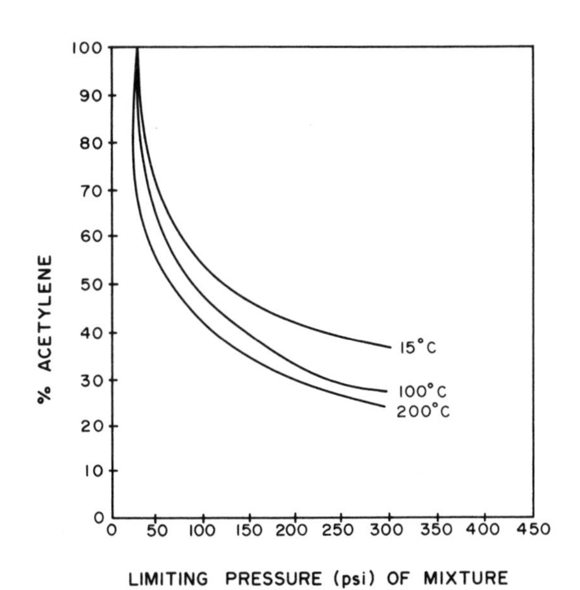

Fig. 1-10 Limiting decomposition pressure for acetylene-nitrogen mixture.

1.4.10 General Principles for Using Acetylene in Chemical Operations

Although acetylene is recognized as a high-energy, hazardous material, it has been handled successfully in commercial operations throughout the world for at least 50 years. The large scale manufacture of butyn-1, 4-diol from acetylene and formaldehyde by companies such as BASF, GAF, and Du Pont show that acetylene can be handled safely under pressure. Only broad principles for safe use of acetylene can be given here, since each technology case will be different, depending upon the reactants, reaction conditions, solvents, properties and stability of the products and by-products formed, together with the method required for product isolation and waste disposal. The following are broad generalizations applicable to either research, pilot plant or production situations.

Pure Acetylene The undiluted gas should be handled and reacted below 15 psig if a vapor phase is present. If transported in pipes of greater than one inch (25 mm) diameter, the pipes should be packed with bundles of smaller pipes. Undiluted acetylene at higher pressures (200-300 psig) should be stored in pressure vessels packed with Raschig rings prior to being added to the reactor.

Acetylene and Inert Diluents If acetylene is diluted to the 50% level (by volume) with inert gases such as nitrogen, carbon dioxide, or methane, its stability at higher C_2H_2 pressures (6-12 atm) is increased. As an inert diluent, nitrogen is generally preferred, although it is less effective than CH_4 and CO_2, ranked in order of effectiveness.

Much batch experimental and pilot work has been carried out over the years throughout the world using acetylene, inert atmospheres and polar solvents in various acetylene reactions (ethynylation, vinylation, cyclization) without reported explosions or detonations. A safe method of carrying out commercial acetylene-based reactions is to operate under continuous conditions and to utilize a solvent system under pressure, where no gaseous phase is ever present. This principal is utilized in butyn-1, 4-diol and propargyl alcohol (Reppe technology) production where acetylene and aqueous formaldehyde are reacted in a complete liquid phase (see II Sec. 2.1-2.4).

Safety Considerations All acetylene chemical processing plants have specialized and somewhat common requirements for handling, reacting, and recycling acetylene. Acetylene must be compressed slowly and the heat removed by cooling. Often, by the use of suitable solvents at low temperatures (-40 to -80°C), it is not necessary to compress acetylene beyond 15 psig. However, acetylene is compressed in the manufacture of butyndiol to at least 75 psig (5 atm) and reacted at 90-130°C. In small, barricaded experimental reactors, acetylene has been compressed to 350-400 psig with no decomposition or explosions noted [42].

The flammability range for acetylene-air mixtures is 3 to 80%, and great care must be exercised that air or oxygen are not accidently introduced into the reaction system. All process lines, fittings, etc., should be made of stainless steel or rust-resistant iron, and copper or silver. Other heavy metals should be avoided. These metals can form unstable metal acetylides which can initiate acetylene decomposition.

The propagation of a decomposition flame via deflagration can be prevented by the use of flame arrestors, packing tubes or pressure vessels with Raschig rings (ceramic or stainless steel), or by packing free space with bundles of small-diameter tubing. The reaction system used should also have effective blow-out disks and adequate vent lines, free from right-angle turns. The most authoritative reference source for all manufacturing aspects of acetylene still remains Miller's treatise [11].

1.5 ACETYLENE ECONOMICS

Any present-day economic discussion concerning acetylene [52-54] relates mainly to the petrochemical-derived product. Calcium carbide acetylene has been almost completely replaced, and no new capacity has been added in recent years. One of the few sites where it is still produced is at Louisville, Kentucky (AIRCO-BOC).

During 1966 the total production of acetylene for use in chemicals manufacture was just over one billion pounds, with the hydrocarbon-derived material amounting to 715 million lb and calcium carbide acetylene rated at 356 million lb [52]. Total U.S. capacity in 1968 for "carbide" acetylene was estimated at 610 million lb, illustrating the large difference between capacity and actual demand.

1.5.1 Manufacturing Cost and Selling Price

The manufacturing cost [53-55] of petrochemical acetylene is very difficult to assess accurately, since it is based mainly on "in-house"

use and is dependent upon the value assigned to feedstocks at the refinery and the final value placed on end products and by-products. The case of acetylene and ethylene in vinyl chloride production is a typical example. Further, the ever-present specter of inflation, feedstock shortages, and competitive use of hydrocarbons for industrial and home fuel use make acetylene and other related product costs an ever changing target.

In the period 1960-1970, manufacturing costs for petro-acetylene were generally quoted at 5-10 ¢/lb, depending on the process and the accounting practices used for other products and by-products. The selling price (contract) averaged 7-12 ¢/lb up to 1970, while by 1973 pipeline acetylene had increased to an average of 14 ¢/lb and by 1976 it ranged from 14-20 cents for larger volume users [9] to over 30 ¢/lb for smaller users. An average price for ethylene during 1970 was 3-5 ¢/lb, while its 1976 price was about 8 ¢/lb. Calcium carbide acetylene during the period 1965-1970 sold for 12-15 ¢/lb and was generally acknowledged to be slightly more expensive to produce than petro-acetylene. During the period 1978-1979 the price of acetylene to larger users was probably in the 30- to 50-¢/lb range. In the period 1980-1982, it is possible that petro-acetylene for chemical use may rise to 75 ¢/lb. The rapidly escalating price of oil, cracking feedstocks, and natural gas, together with the continuing shutdown of petrochemical acetylene plants, makes a high acetylene price possible.

Calcium carbide (79-83% pure) was selling in 1979 for $285/t or about 14 ¢/lb. The materials cost for acetylene liberated from calcium carbide at a 100% yield (0.31 lb acetylene per pound calcium carbide) is 46 ¢/lb. Since acetylene generation from carbide is a simple step and is not labor intensive, production costs of acetylene are probably 50-55 ¢/lb to a large volume supplier such as AIRCO at their Louisville plant.

The calcium carbide process, based on abundant limestone and coal (coke), is not subject to the unstable price, supply, and political problems of oil. It is possible that carbide acetylene can already compete with petrochemical acetylene, and in the years ahead there may be a clear-cut advantage. In Sec. 1.1, 1.2, 1.3.2, and 1.7.2 the future potential of coal-based acetylene as a raw material source for commodity chemicals manufacture is discussed.

1.5.2 Earlier Growth and Uses

During the period 1961-1970 the growth of petrochemical acetylene averaged about 16%/year. By 1970, however, it was apparent that acetylene use for monomers such as vinyl chloride and vinyl acetate had begun to decline rapidly due to alternative processes based on ethylene. Total annual capacity of acetylene during 1971 was 468 million lb with demand equal to capacity [55]. By 1976 petrochemical acetylene capacity had undergone a further decline to 419 million lb, while demand had decreased to a level of 325 million lb. The decline continued into 1979, with capacity at 406 million lb and demand at 269 million lb.

For 1976 the following breakdown of chemical uses for acetylene was reported [55]: vinyl chloride, 31%; acrylates, 26%; vinyl acetate, 18%; acetylenic chemicals (Reppe types, acetylenic alcohols and glycols), 9%; tetrahydrofuran (THF), 8%; chlorinated solvents 4%; other uses, 4%. Besides its decline in vinyl chloride and vinyl acetate production, acetylene has now been replaced by propylene as the preferred feedstock for acrylates and by butadiene for chloroprene (neoprene rubber) production. In contrast, the use of acetylene for Reppe chemicals and THF along with lower volume specialty acetylenics is expected to continue growing over the years.

1.5.3 Producers and Capacity

Producers of acetylene for 1976 and 1979 are listed below. Most of these facilities also produce ethylene as a coproduct. AIRCO and Hoffman-La Roche are the only producer and user, respectively, of calcium carbide acetylene. The preferred feedstock for petrochemical acetylene is natural gas.

	Acetylene Capacity (MM lb/yr)	
Producer	1976	1979
AIRCO, Calvert City, KY		50
Dow Chemical, Freeport, TX	15	16
Hoffman-La Roche, Nutley, NJ		5
Monochem, Geismar, LA	180	180
Rohm and Haas, Deer Park, TX	55	0
Tenneco, Houston, TX	100	100
Union Carbide, Ponce, PR	12	12
Union Carbide, Seadrift, TX	20	15
Union Carbide, Taft, LA	12	10
Union Carbide, Texas City, TX	25	16
Union Carbide, Torrance, CA		2
	419	406

Production figures for calcium carbide acetylene in 1976 are not reported but it is likely they are less than one-fifth of total petrochemical production. The principal carbide producers of acetylene are AIRCO, with a capacity of 300 million lb/yr at Louisville, and the former Ferro-Alloys Division of Union Carbide at Ashtabula, Ohio, with a capacity of 75 million lb/yr.

The decline of carbide acetylene became significant in 1972. Large production facilities at Niagara Falls (Du Pont, 75 million lb/yr) and South Charleston, West Virginia (Union Carbide, 60 million lb/yr) were shut down. Shortly after, carbide acetylene production at Calvert City (Air Reduction) was halted and calcium carbide from AIRCO's Louisville plant was used to generate acetylene at the Calvert City site for the chemical needs of Air Products and Chemicals (APCI) and General Aniline and Film (GAF) at these "across the fence" locations. AIRCO's Louisville carbide plant prior to 1971 supplied most of its acetylene to Du Pont (at the same location) for the production of chloroprene and polychloroprene (neoprene rubber). This process was replaced by new technology based on the use of butadiene in place of acetylene (see Sec. 1.6.7). Some acetylene, however, is still supplied to Du Pont for the production of vinyl fluoride and polyvinyl fluoride.

It is still possible to use acetylene for vinyl chloride and vinyl acetate production if it is used to augment the ethylene-based processes and improve overall process economics. However, in the case of acrylate production it is now certain that propylene is the preferred feedstock, even though the acetylene-carbon monoxide-nickel carbonyl route has been dominant for over 30 years (see Sec. 1.6.2 and 1.6.3). Rohm and Haas, the major user of acetylene for acrylate production at Deer Park, Texas has shut down its acetylene-based facility. Although it is on a stand-by basis, there are no near-term plans of using it, unless the price of propylene should escalate markedly.

1.6 COMMODITY CHEMICALS FROM ACETYLENE AND OTHER HYDROCARBONS

Figure 1-11 shows the large-volume chemicals that can be produced from acetylene and other hydrocarbons. The raw material in parentheses denotes the currently favored feedstock, while below it summarize the technology used. The raw materials of choice currently are ethylene, propylene, and butadiene. Acetylene black is listed with these commodity chemicals for convenience,

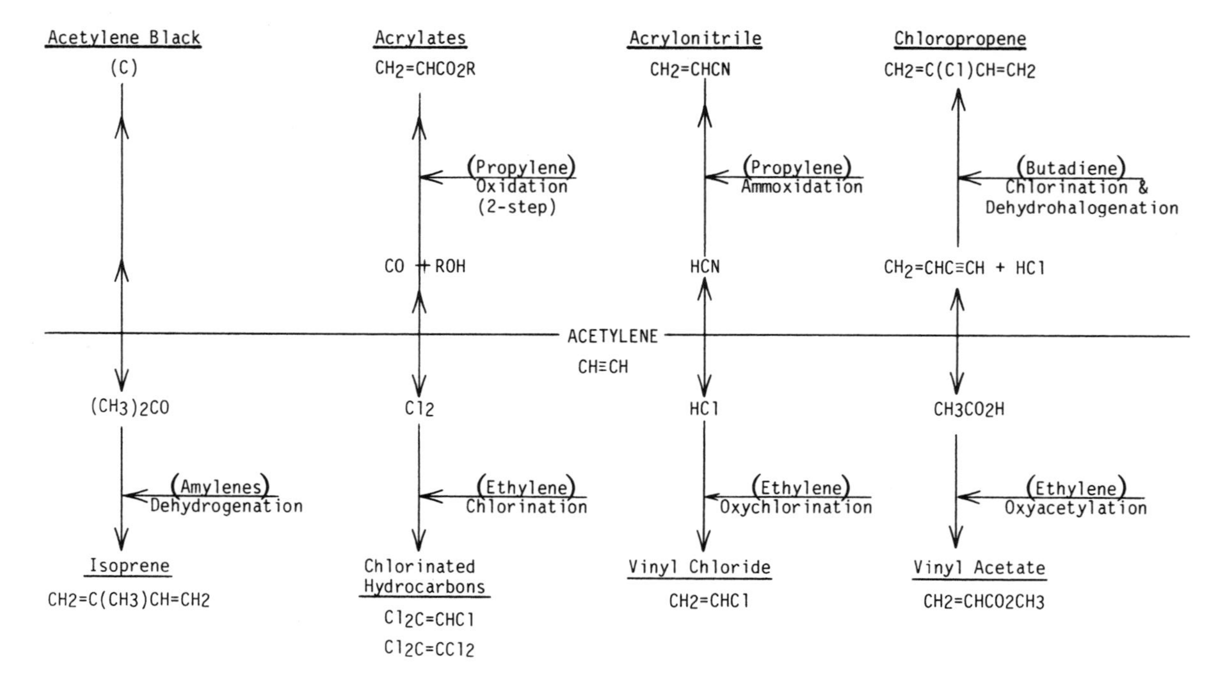

Figure 1-11 Commodity chemicals from acetylene and other hydrocarbons.

although it is a specialty product and of considerably lower production volume.

In the sections that follow, processes based on acetylene and other hydrocarbons for commodity chemicals are discussed. In Chap. 2, the growing and diverse areas of large-volume and specialty acetylenic chemicals is discussed with respect to applications and technology. In Sec. 1.1, 1.3.2, and 1.7.2 reasons are presented for the eventual renaissance of acetylene as a raw material for the production of such products as acrylates, chloroprene (neoprene), chlorinated hydrocarbons, vinyl chloride, and vinyl acetate, based on the use of by-product acetylene and coal-based acetylene.

1.6.1 Acetylene Black

This unique, highly conductive form of carbon is prepared by the exothermic decomposition of acetylene at 800°C or above. The technology is old and dates back to before the 1900s [56]. Acetylene black can be prepared by the simple thermal (800°C) decomposition of acetylene into its elements.

$$C_2H_2 \xrightarrow{\Delta} 2C + H_2$$

The reaction is highly exothermic (∿55,000 cal/g), and the heat evolved is used to maintain pyrolysis temperature in the production furnace. The continuous thermal process has been practiced commercially in Europe, Canada, and the Far East [57]. In contrast, the thermal decomposition of olefins, alkanes, and aromatic hydrocarbons to carbon is an endothermic reaction.

Acetylene black has often been an undesirable byproduct of electric arc technology for the production of acetylene from hydrocarbons. An arc process for its manufacture has been proposed [58].

The first (1860s) commercial acetylene black process involved partial combustion of acetylene to produce a carbon similar to the petrochemical channel or gas blacks [59]. The process that is now favored is the continuous explosion of acetylene at low pressure (1-2 atm) using an electric spark to initiate the reaction. The process is brought to decomposition temperature (800-1,000°C) by initial combustion of acetylene with air, and then shutting off the air supply. The resulting reaction proceeds by both explosion (electric discharge) and thermal conversion to carbon and hydrogen [60].

The reactor is generally a metal retort lined with refractory brick, and is preferably water cooled to control the highly exothermic reaction. The carbon is discharged from the bottom of the retort and is then compressed to a bulk density of 13 lb/ft^3. Acetylene black initially formed is a very light, fluffy mass having a bulk density of only 1.2 lb/ft^3.

The principal use for acetylene black is the construction of dry cell batteries, where its desirable properties of high electrical conductivity, high purity, small particle size, low moisture absorption, chain-cyclic structure, compatibility, and reactivity with manganese dioxide make it unique and valuable. The benefits realized by its use are lower cell resistance, higher capacity, and longer shelf life.

The use of acetylene black in rubber compounding and in plastics and coatings applications is limited mainly to conductive rubber, plastic sheeting, and molded products. Its higher price makes it too expensive to compete successfully with petrochemical-based carbon in typical large-volume compounding uses.

The United States market for acetylene black in 1967 was about 10 million lb, which grew to about 13 million lb by 1972. Continued growth in specialty areas is expected.

The principal North American producers are Shawinigan (Canada) and Union Carbide (Ashtabula, Ohio). The latter company, during 1966 [9] was reputed to use about 3 million lb/yr of acetylene, from calcium carbide, for acetylene black production. This plant was producing at the rate of 8 million lb/yr in 1975. Also, Union Carbide is operating a relatively new acetylene black production facility at Ponce, Puerto Rico.

Gulf announced [131] it would manufacture acetylene black in 1979 at its Cedar Bayou, Texas site using by-product acetylene from its olefins production plant. The capacity of the new facility is rated at 15-20 million lb/yr. The increasing use of dry cell batteries in children's toys and other recreation devices, together with other conducting and antistatic applications, makes acetylene black an attractive and profitable larger volume specialty product.

1.6.2 Acrylate Esters and Acrylic Acid

Approximately one-third of the world's production of acrylate esters and acrylic acid is still manufactured by the acetylene-carbon monoxide route, sired by Julius Reppe and coworkers as part of a varied array of acetylenic chemicals and derivatives. However, the sands of time are running out on this technology, the ever-rising costs of petrochemical acetylene, and the rising costs of special plants needed to use, produce, and recycle the highly toxic nickel carbonyl catalyst used for acrylate production being the primary reasons. In an era where toxicity and pollution are prime considerations in an ever stricter legislative atmosphere, the Reppe carbonylation route has become unattractive. Older amortized facilities will probably continue to be used or retained on a stand-by basis.

The new oxidation route based on the two-stage oxidation of propylene to acrolein and acrylic acid, respectively, followed by esterification, has made significant inroads against the Reppe process. All new plants since the early 1970s have been based on propylene.

Acetylene–Carbon Monoxide Process The early Reppe carbonylation plants (in the 1940s) were operated at or near atmospheric pressure, using a stoichiometric amount of nickel carbonyl with respect to acetylene [48, 61, 62].

(a) $4C_2H_2 + Ni(CO)_4 + 4C_2H_5OH + 2HCl \longrightarrow CH_2{=}CHCO_2C_2H_5 + H_2 + NiCl_2$

The reaction gave essentially quantitative yields of acrylate esters or acrylic acid under mild temperature (30-40°C) conditions, by using either alcohols or water as coreactants. An important asset of the C_2H_2-CO route is the flexibility of producing a variety of acrylates in the same system.

Since it was known [62, 63] that the carbonylation of acetylene could also be carried out catalytically with nickel salts ($NiBr_2$) under pressure (30 atm) at higher temperatures (180-200°C), it was not long before it was recognized [64, 65] that nickel carbonyl could be used in catalytic amount in the presence of excess carbon monoxide. The process defined by the stoichiometry shown below could be operated at conditions approximating the stoichiometric process.

(b) $C_2H_2 + 0.8\ CO + 0.05\ Ni(CO)_4 + 0.1\ HCl + C_2H_5OH \longrightarrow CH_2{=}CHCO_2C_2H_5 + 0.05\ NiCl_2 + 0.05\ H_2$

An important advantage of this method was that it used much less of the highly toxic nickel carbonyl ($NiCO_4$), making manufacture and recycling of $Ni(CO)_4$ from $NiCl_2$ less of a problem. This semicatalytic process has been in use since 1952 in the Rohm and Haas production facility at Deer Park, Texas.

Figure 1-12 shows a flow diagram for the C_2H_2-CO-$Ni(CO)_4$ process [64]. The semicatalytic process is initiated by the stoichiometric reaction (a), but once an excess of carbon monoxide is introduced to the system, the catalytic reaction predominates, with only 20 to 40% of the CO being supplied by the nickel carbonyl. The reaction temperature approximates the 40°C range of the stoichiometric process, but it is likely the process is run under low (15 psig) CO-C_2H_2 pressures.

The reactants C_2H_2, CO, $Ni(CO)_4$, HCl gas, and alcohol (methanol or ethanol) are fed continuously under the surface to a stainless steel, stirred-tank reactor equipped with a recycle-heat-exchange loop for controlling reaction temperature. The following average stoichiometry is used through the reactor, which approximates reaction (b). Accurate feed flows are required for optimum results.

Reactant Mixture		Mole Ratio
(A)	(B)	(A/B)
C_2H_2	CO	1.0-1.1
Alcohol	CO	1.1-3.0
HCl	$Ni(CO)_4$	0.8-1.0
C_2H_2	$Ni(CO)_4$	20

The emerging reactor stream is processed to monomer, grade acrylate ester, via two main unit operations, separation and purification, as shown in Fig. 1-12.

Prior to final venting to the atmosphere, inert gases, vapor impurities, and traces of alcohol and nickel carbonyl are scrubbed with alcohol to recover usable materials and avoid air contamination. The vent gas is then flared and organic liquid wastes burned in a special incinerator for recovery of fuel value to the process.

Although the carbonylation process has the potential to directly produce a variety of acrylate esters by the use of different alcohols, in commercial practice the methyl and ethyl esters are mainly produced. The higher esters, such as the butyl and 2-ethylhexylacrylates, are produced by ester interchange reaction with methyl acrylate.

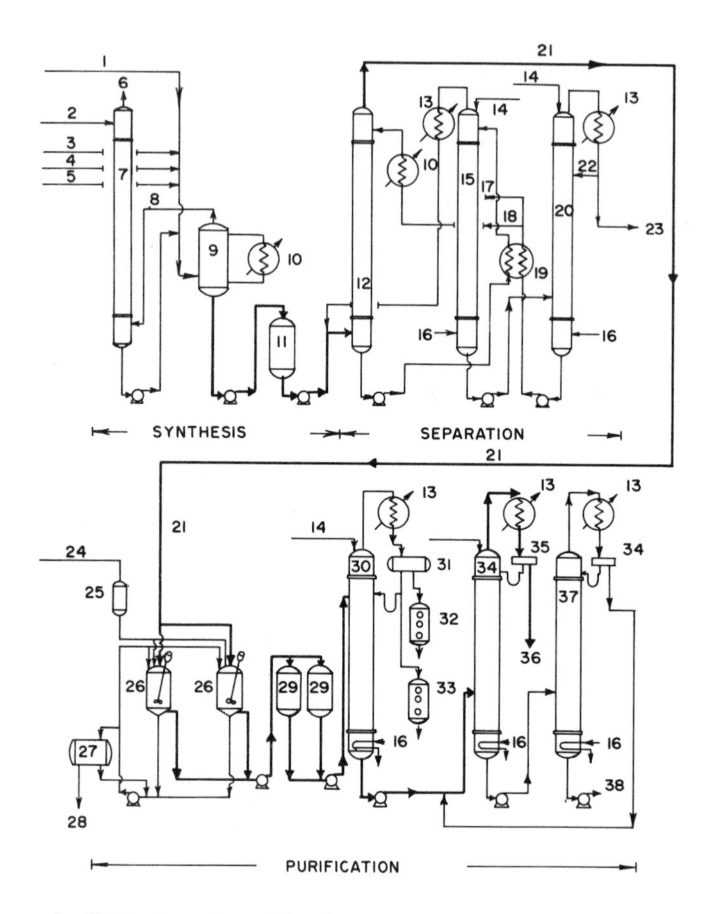

1. Nickel carbonyl feed
2. Alcohol feed
3. Carbon monoxide feed
4. Hydrogen chloride feed
5. Acetylene feed
6. Vent
7. Vent gas scrubber column
8. Vent to (7)
9. Reaction kettle
10. Water cooling
11. Holding tank
12. Extraction column
13. Water condenser
14. Inhibitor feed
15. Rectifier
16. Steam
17. Excess brine to nickel carbonyl plant
18. Recovered brine
19. Heat exchanger
20. Brine stripping column
21. Wash crude to rerun area (26)
22. Reflux
23. To alcohol dehydration
24. Soda ash solution
25. Soda ash measuring tank
26. Twin crude neutralization tanks
27. Holding tanks
28. Soda ash solution to sewer
29. Twin crude feed tanks

Production and Recycling of Nickel Carbonyl Nickel chloride solution [64] composed of stripped process brine and fresh nickel chloride solution is introduced via high-pressure injection pumps to a stirred pressure reactor concurrently with excess sodium hydroxide solution. Carbon monoxide is then added through the bottom of the reactor via a four-stage compressor. The conversion to nickel carbonyl takes place continuously above 100°C and at pressures of about 1500 psig. The yield and selectivity to nickel carbonyl is high, probably over 95%.

The nickel carbonyl (bp 43°C) is condensed in the product receiver and stored under refrigeration under a carbon monoxide atmosphere. Excess CO is recycled to the reactor section. The remaining liquid slurry is processed as waste. Inert gas buildup is taken care of for nickel carbonyl content by passage through a decontamination furnace. Other volatiles also are flared.

1.6.3 Propylene Oxidation Process for Acrylic Acid

The primary advantages of the propylene route over acetylene-based technology are the cheapness of propylene (10-12¢/lb) versus acetylene (15-25¢/lb) and a much lower toxicity and pollution problem. Although the oxidation process appears more complex than the acetylene-CO route, it has been made more efficient by the erection of very large (>200 MM lb/yr) plants and the development of improved oxidation catalysts. These factors cancel out the higher yield (∿95-100%) of the acetylenic route over the propylene to acrylate ester yields of about 60 to 70%.

The oxidation of propylene to acrylic acid is best carried out in two separate reactors (Ref. 18, pp. 337-342; Ref. 66, 67) connected in series as shown in Fig. 1.13. In reactor (A), propylene in the presence of excess air and steam is oxidized mainly to acrolein, together with varying amounts of acrylic acid, CO_2, CO,

---

30. First stage rerun column
31. Gravity separator
32. Water receiver
33. Light ends receiver
34. Second stage rerun column
35. Flow splitter
36. Finished monomer to storage
37. Third stage rerun column
38. Residue to disposal

Figure 1-12 Acetylene-carbon monoxide-nickel carbonyl route to acrylates.

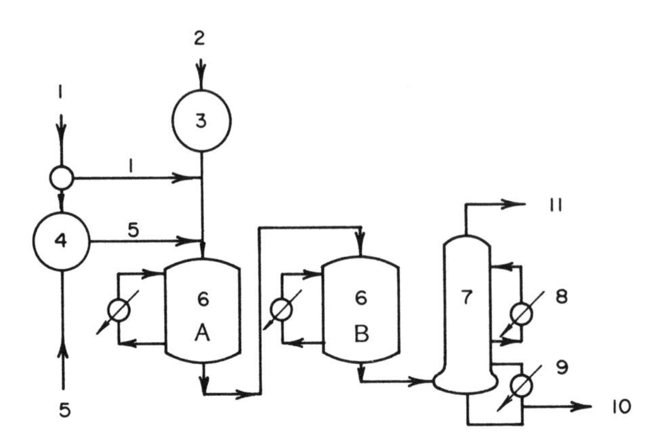

1. Steam feed
2. Air feed
3. Air compressor
4. Vaporizer
5. Propylene feed
6. Reactors A and B with fused salt heat exchangers
7. Aqueous absorber
8. Intercooling loop
9. Bottoms cooling loop
10. Dilute acrylic acid for for purification
11. Exit gas to flare or incinerator

Figure 1-13 Propylene oxidation to acrolein and acrylic acid.

and tars formed as by-products. In reactor (B), the acrolein rich mixture is further oxidized to acrylic acid.

$$CH_2CH{=}CH_2 \xrightarrow{(A)} \underset{\text{acrolein}}{CH_2{=}CHCHO} \xrightarrow{(B)} \underset{\text{acrylic acid}}{CH_2{=}CHCO_2H}$$

The effluent from reactor (B) contains besides acrylic acid (20% concentration), acetic acid, acetone, acrolein, carbon monoxide, carbon dioxide, and water. The aqueous solution is extracted (butyl acetate), concentrated, and purified by vacuum fractionation to glacial acrylic acid, which is then esterified via the standard alcohol-sulfuric acid method.

$$CH_2{=}CHCO_2H + CH_3OH \xrightarrow{\text{concd } H_2SO_4} CH_2{=}CHCO_2CH_3 + H_2O$$

The overall yield to acrylic acid averages 67 to 83% [18,67] depending upon the catalyst, conditions, and the process used. The recovery of acrylic acid from the aqueous reactor effluent averages over 95%. The overall yield to acrylate ester is about 60 to 70% of the theoretical yield based on propylene [67].

The reactors (A) and (B) are fixed-bed, shell, and tube [18,67] units. The tubes packed with catalyst are about 10 to 15 ft in length and about 1 in. inside diameter. A fused salt mixture circulates through the shell to control the exothermic oxidation, and heat is removed by an external heat exchanger loop. The first reactor is controlled at an average peak temperature of 330-450°C, utilizing a preheated feed comprised of about 5% propylene, 40% steam, and about 55% air. Most of the reaction exotherm is due to the formation of carbon dioxide and carbon monoxide compared with acrolein formation.

Reactor (B) operates at a somewhat lower peak reactor temperature (275-365°C) and further oxidizes the acrolein containing gaseous effluent from the first reactor to acrylic acid. The second-stage gaseous effluent is fed to the bottom of a water-acrylic acid absorber, where it is cooled to below 80°C by a combination of water entering the top of the column and absorber liquid being cooled by the external heat exchanger loop. The aqueous effluent averages about 25% acrylic acid. The residual off-gas is either flared or burned in a furnace to an acceptable composition before being vented to the atmosphere.

The aqueous acrylic acid is extracted with a solvent or mixture comprised of butyl acetate, diisobutyl ketone, toluene or xylene, to recover acrylic acid and by-products such as acetic acid, acetone, and acrolein. The acrylic acid is readily separated in high purity (98-99%) by a standard series of continuous distillation columns, which have been described in detail [18].

1.6.4 Acrylonitrile: Acetylene-Hydrogen Cyanide Route

The vapor-phase reaction of acetylene and hydrogen cyanide over an alkali metal or alkaline earth ($Ca(CN)_2$] catalyst at 400 to 600°C belongs to the realm of extinct technology, not likely to be used again. During the early 1960s over 60% of domestic acrylonitrile production was by this route [68]. Prior to the vapor-phase route, a liquid-phase process used by American Cyanamid Corporation also was practiced from the 1940s to the 1960s. However, with the advent of SOHIO technology [69, 70] via the reaction of propylene, air, and ammonia to directly form acrylonitrile, all other processes were quickly phased out. Some new vapor-phase C_2H_2-HCN plants only several years into production suffered this fate. Acetylene-based and other technology has been summarized by Miller (Ref. 11, Vol. 2, pp. 156-183) and Othmer [13, 18]. Besides unfavorable economics, the C_2H_2-HCN route has the further disadvantage of using highly toxic hydrogen cyanide, which requires special handling, recycling and disposal procedures.

During 1970 Monsanto announced [71] it was closing down all petrochemical acetylene production (100 million lb/yr) at Texas City, together with such integrated facilities as acrylonitrile (130 million lb/yr) and vinyl acetate (80 million lb/yr). In contrast, their propylene-based acrylonitrile plant at Alvin, Texas was expanded to 370 million-lb capacity to offset the Texas City unit. During this period American Cyanamid shut its plant also and converted to SOHIO technology.

Propylene Ammoxidation (SOHIO) Technology The acrylonitrile process developed by SOHIO [72, 73] has evolved into one of the more important technologies of the chemical industry and has been licensed and practiced worldwide. It is today the sole process used for acrylonitrile production throughout the world.

The catalyst originally developed was of great importance to the success of this process. Prior to its advent the ammoxidation reaction proceeded in poor yield with much by-product formation. Molybdenum was found to be a key element for solving the unique reaction catalysis required for this ammoxidation reaction:

$$CH_3CH{=}CH_2 + \frac{3}{2}O_2 + NH_3 \longrightarrow CH_2{=}CHCN + 3H_2O$$

The first successful catalyst [69] employed by SOHIO was a bismuth-phosphorous-molybdenum-oxygen composition (bismuth phosphomolybdate) deposited on colloidal silica gel. Later catalyst compositions (see the following section) involved the use of a variety of other elements, such as vanadium, tellurium, cerium, selenium, copper, antimony, iron, tin, and uranium [74], most often formulated with molybdenum.

Ammoxidation Processes The earlier processes utilized fixed-bed tube and shell reactors in which a steam diluent was also used to control the highly exothermic reaction. Most modern plants today utilize a fluidized-bed reaction system, which provides excellent temperature control, high conversions, and a contact time of several seconds. The process generates its own steam diluent, and this steam source is used in downstream process operations involved with product isolation. Some typical ammoxidation processes are tabulated here:

Process	Catalyst	Reaction Temperature (°C)	Reactor
SOHIO [69]	Bi-P-MO	400-500	Fixed and fluid bed
SNAM [75]	MO-Vo-Bi	440-470	Fixed bed
Montedison-UOP [76]	MO-Te-Ce	240-460	Fluid bed
BP (Distillers)	(1)Se-CuO	–	Two-reactor system
Ugine [77]	(2)MoO_3		

The BP process involves partial oxidation of propylene to acrolein with a Se-CuO catalyst, followed by reaction (ammoxidation) of the acrolein with air and ammonia. The yield to acrylonitrile claimed in the second step is high (>90%). BP has also developed an antimony-tin-ferric oxide catalyst for the direct conversion of propylene to acrylonitrile [78]. Most acrylonitrile processes operate at low pressures (3-30 psig) to effect heat recovery and utilization in the process.

The Montedison-UOP acrylonitrile process [76,79] is typical of new technology comprising both a new, highly efficient catalyst and a fluidized-bed reactor. Figure 1-14 shows the flow diagrams for the reactor and product isolation reactions. Propylene, air, and ammonia are fed to the fluidized-bed reactor in close-to-stoichiometric ratio, using a new catalyst (MO-Te-Ce) composition. The highly exothermic reaction (123-150 kcal/mol) results in over 95% of the propylene being converted to products, with the selectivity to acrylonitrile being over 80%.

Typical by-products of all ammoxidation processes are acetonitrile, hydrogen cyanide, CO_2, CO, and a minor amount of carbonyl compounds. Acetonitrile is sold as a specialty (highly polar) solvent useful in plastics applications, while hydrogen cyanide can be employed in the acetone cyanohydrin process for the production of methacrylate esters. However, it is also likely that much of these by-products are also incinerated or flared, since the markets for these by-products are limited.

The high-pressure steam generated in the ammoxidation step is used to control the reaction temperature and provide steam to operate the compression system [79]. Cooling the reactor effluent

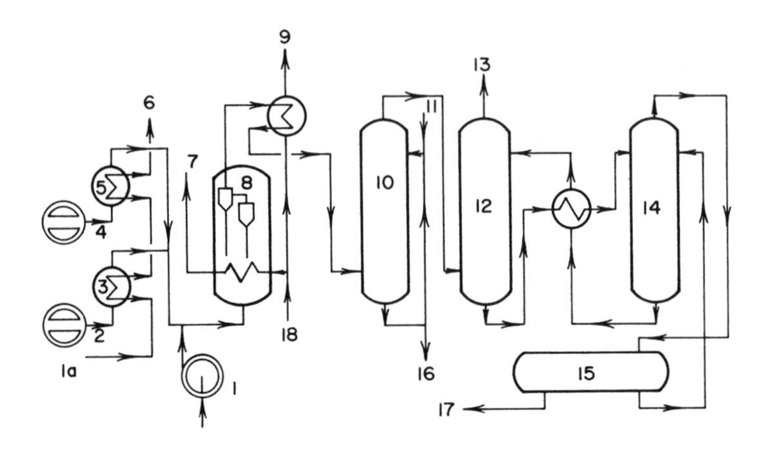

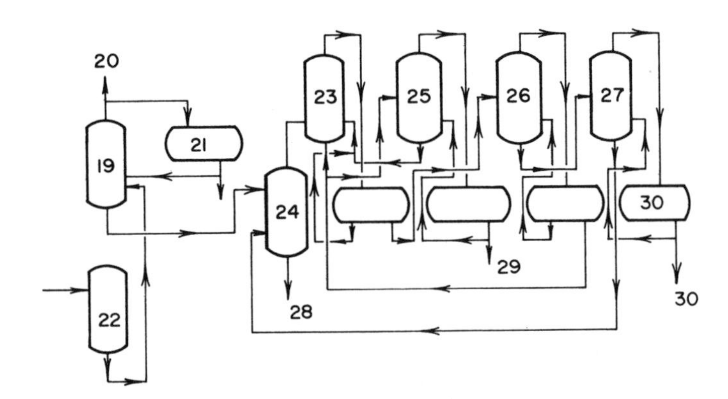

1. Air feed and air blower
1a. Glycol-water solution
2. Ammonia feed
3. Ammonia evaporator
4. Propylene feed
5. Propylene evaporator
6. For refrigeration cooling
7. High pressure steam
8. Fluidized bed reactor
9. MP steam
10. Quench neutralizer
11. Sulfuric acid
12. Absorber
13. Vent/fuel gas
14. Stripper
15. Stripper receiver
16. Ammonium sulfate solution
17. To intermediate storage
18. Boiler feed water
19. HCN separation
20. HCN gas
21. Liquid HCN receiver
22. Intermediate storage
23. Extractive distillation-high boilers removal
24. High boilers receiver
25. Acetonitrile stripper
26. Dehydration column

provides additional steam used for downstream separation and purification units. The quenched, neutralized reactor effluent is processed through the multiple steps shown in Fig. 1-14.

Acrylonitrile and acetonitrile, having very close boiling points, are separated from each other by extractive distillation with water. An important feature of the separation-purification units (Fig. 1-14) is the operation of some of the columns at different pressures to realize more efficient heat recovery. The acetonitrile-water extract is concentrated to 60-70%, from whence it can be readily purified if desired. The acrylonitrile obtained by final distillation is of high purity (99.9%).

The recovery and disposal of process by-products, such as hydrogen cyanide, acetonitrile, carbon monoxide, carbon dioxide, and nitrogen, together with process economics has been discussed by Pujado and coworkers [79]. For the manufacture of 100,000 t/yr of acrylonitrile, the annual raw materials consumption of propylene and ammonia is 118,000 and 48,000 t, respectively. The Montedison-UOP process is claimed to be one of the most efficient of the new ammoxidation processes.

1.6.5 Acetaldehyde

The manufacture of acetaldehyde (Ref. 11, Vol. 2, pp. 134-147) dates back to circa 1916 in Germany and the early growth of the calcium carbide acetylene industry. Prior to 1940, the acetylene-based route was the principal manufacturing process. The principal use for acetaldehyde then, as today, was the manufacture of acetic acid and its anhydride. The early process involved the hydration of acetylene at 70-100°C in 18-25% sulfuric acid, using mercury salts as catalysts in a continuous tower-type reactor.

$$CH{\equiv}CH + H_2O \longrightarrow CH_3CHO$$

By 1969 no acetaldehyde was manufactured in the United States by this method, except a small amount (less than 5% of total production) derived as by-product from chloroprene, acrylonitrile, and vinyl acetate processes. The favored commercial routes now

---

27. Acrylonitrile rerun column
28. Heavy residue to incineration
29. Crude acetonitrile
30. High purity acrylonitrile

Figure 1-14 Montedison-UOP acrylonitrile process.

are the oxidation of ethylene ($PdCl_2$ catalyst), saturated hydrocarbons (butane), the oxidation of ethanol (silver catalyst) (Ref. 13, pp. 84-89). Unless a major technological breakthrough is attained to obtain cheap acetylene, or oil-based feedstocks become too expensive and scarce to use, it is doubtful that the hydration route will be used again.

1.6.6 Chloroprene (2-Chloro-1,3-butadiene)

The technology for the production of chloroprene [80] and its derived homopolymer, neoprene, dates back to the 1930s. Du Pont, utilizing calcium carbide acetylene at Louisville from the nearby Air Reduction Company site, practiced the two-step process based on acetylene for over 30 years. The process is based on the discovery of the dimerization of acetylene by J. Nieuwland and of chloroprene by W. Carothers [80].

$$2CH{\equiv}CH \xrightarrow{CuCl\text{-}NH_4Cl} CH_2{=}CH\text{-}C{\equiv}CH$$

vinyl acetylene

$$CH_2{=}CH\text{-}C{\equiv}CH + HCl \longrightarrow CH_2{=}CH\text{-}\underset{}{\overset{Cl}{\overset{|}{C}}}{=}CH_2$$

chloroprene

The practice and control of this hazardous technology enabled Du Pont to become the dominant U.S. producer of neoprene rubber. Today the acetylene-based route has been supplanted by a cheaper process based on the selective chlorination of butadiene.

Vinyl acetylene [81,82] is made continuously by dimerizing acetylene in a 55% solution of CuCl-$2NH_4Cl$ also containing copper powder and HCl at about pH 1. Distilled vinyl acetylene is then reacted continuously with HCl in a liquid phase system [83,84] also containing cuprous chloride and copper to yield 4-chloro-1,2-butadiene (isochloroprene) via 1,4 addition, which rearranges in the reaction medium to chloroprene. The principal use for chloroprene is the production of the chemically resistant rubber polychloroprene (neoprene).

Figure 1-15 is a schematic flow diagram for the integrated production of monovinyl acetylene (MVA), chloroprene, and polychloroprene rubber. Similar separate flow diagrams for MVA and chloroprene production have been described [85,86].

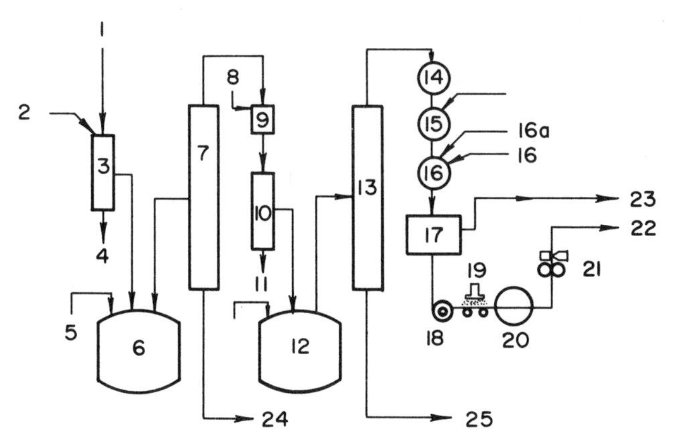

1. Acetylene feed
2. Scrubbing solution
3. Scrubber
4. Scrubbing solution effluent
5. Cuprous chloride solution feed
6. Monovinylacetylene (MVA)
7. MVA fractionation tower
8. Steam
9. Vaporizer
10. MVA scrubber
11. Acid water effluent
12. Chloroprene reactor
13. Chloroprene fractionation tower
14. Water weigh tank
15. Emulsion tank
16. Catalyst feed and polymerization tank

16a. Stabilizer

17. Blending tank
18. Coagulating roll
19. Wash belt
20. Dryer
21. Cutting and roping machines
22. Dry neoprene
23. neoprene latex
24. Divinylacetylene and impurities
25. Waste products

Figure 1-15 Neoprene production from chloroprene and vinyl acetylene.

This process is considered hazardous, primarily due to such reactants, intermediates, and by-products as acetylene, monovinyl acetylene (MVA), chloroprene, and divinyl acetylene. Since acetylene is handled at near-atmospheric pressure, its hazards, in the absence of initiation from the other process intermediates, are minor. However, the remaining products form unstable, explosive peroxides in air, particularly divinyl acetylene, and special process procedures are required to operate this process safely.

In the early 1960s an explosion of unknown origin leveled the Du Pont Louisville facility and posed serious flame-heat and detonation hazards to the nearby (across the fence) acetylene producing plant of Air Reduction Company. The Du Pont plant, however, was quickly rebuilt and operated safely until the newer butadiene-based technology replaced the acetylene route.

1.6.7 Non-Acetylene-Based Processes for Chloroprene

In recent years alternate petrochemical routes based on butane and butadiene [87] have been evaluated to replace acetylene. Butadiene is available in multitonnage quantities via the catalytic dehydrogenation of butane, and represents fairly cheap (10¢/lb, 1971; 21¢/lb, 1979), high-purity starting material for the production of chloroprene.

By 1969-1970 Du Pont began operating a new chloroprene plant at La Place, Louisiana that was claimed to have a capacity of about 80 million lb/yr of neoprene rubber. The process [88] used butadiene as feedstock and involved continuous high-temperature (500°C) free-radical chlorination of butadiene. Under these conditions, radical (Cl•) substitution of hydrogen was favored over addition to double bonds, and the isomer problem associated with the lower temperature chlorination process was largely avoided.

$$Cl_2 \longrightarrow 2Cl\cdot$$

$$CH_2{=}CH\text{-}CH{=}CH_2 + 2Cl\cdot \xrightarrow{600°C} CH_2{=}CH\text{-}\underset{}{\overset{CL}{\overset{|}{C}}}{=}CH_2 + HCl$$

The chlorination route still requires careful control of reaction conditions to avoid addition to double bonds and polymerization of butadiene. Also, the handling of chlorine and hydrogen chloride at high temperatures can be troublesome due to special equipment, corrosion, hazards, toxicity, and special process operating procedures. Also, hydrogen chloride must be either oxidized back to chlorine for recycling or else sold as a process credit.

Once the new butadiene chlorination technology had been shown to be viable, Du Pont closed down its acetylene-based plants at Montague, Michigan and Louisville, Kentucky in 1972. At that time both plants utilized about 242 million lb/yr of acetylene. The Louisville plant was subsequently converted to the butadiene route; by 1973 acetylene was no longer used for neoprene production. In 1970 the Petro-Tex Corporation (Houston) also began operating a butadiene-based facility, thereby becoming a second supplier.

The end use for chloroprene is almost exclusively directed to the production of polychloroprene rubber, which has specialty uses in such applications as wire and cable coatings, footwear, industrial belts, and numerous oil and ozone resistant rubber products [89,90]. Besides its uses as a solid and latex elastomer, neoprene in recent years has been increasingly used in flexible-foam applications. Its resistance to oxidation and its flame-retardant properties make it desirable for use in adhesive, tapes, carpet backing, mattresses, and seating cushions. However, the most important markets for neoprene are still varied industrial rubber goods, particularly those used in the automobile and construction industries.

Domestic production of neoprene in 1974 set a record of 135,000 t, but in 1975-1976 averaged only 110,000 t. By 1977, consumption of neoprene was about 167,000 t, showing a modest growth rate (2-3%) based on new developing uses [90].

1.6.8 Isoprene From Acetylene and Acetone

The formation of isoprene via the ethynylation of acetone to 2-methyl-3-butyn-2-ol (MB), followed by its semihydrogenation to 2-methyl-3-buten-2-ol (MBe) and finally dehydration to the desired monomer, has been studied [91-93] and compared with other petrochemical routes [94,95].

$$(CH_3)_2C{=}O + C_2H_2 \longrightarrow (CH_3)_2C(OH){-}C{\equiv}CH \quad (90\text{-}100\%)$$

(MB)

$$MB + H_2 \longrightarrow (CH_3)_2C(OH){-}CH{=}CH_2 \quad (99\%)$$

(MBe)

$$MBe \longrightarrow CH_2{=}C(CH_3){-}CH{=}CH_2 + H_2O \quad (95\text{-}100\%)$$

Although initial economics based on high overall yields and process efficiency appeared attractive for manufacture in the United States, this method ultimately lost out to processes based on either propylene or C_5 hydrocarbons during the period 1960-1970.

However, in Italy a continuous process [96,97] was developed for Ente Nazionale Idocarburi (ENI) by its research subsidiary SNAM Progetti based on the acetone-acetylene route, and a production unit is being operated at Ravenna, Italy [98,99]. The process originally used a catalytic amount of sodium in the ethynylation step, which was subsequently converted to sodium acetylide, which is the true catalyst species. An alternative catalytic ethynylation process using NaOH or KOH in liquid ammonia [93] has been developed and is being used for the commercial production of methyl butynol. The plant at Ravenna operated by ANIC (affiliate of ENI) was designed to supply isoprene for the production of 30,000 t/yr (66 million lb/yr) of cis-1,4-polyisoprene [98]. The complete isoprene process was offered for patent licensing in 1970 [99].

Figure 1-16A is a flow diagram of the continuous acetone-acetylene route for MB production [100]. Figure 1-16 describes the reaction (ethynylation) of acetone and acetylene with a catalytic amount of 50% aqueous KOH in liquid ammonia as reaction media. The yield and selectivity to methyl butynol (MB) is over 95% using a minimum acetylene/acetone mole ratio of 1.5:1.

Acetylene, acetone, ammonia-acetylene (recycle) and KOH catalyst solution are mixed before entering the liquid-phase tubular reactor. The reaction is controlled at 10 to 41°C and 280 to 350 psi. The KOH catalyst in the reactor effluent is neutralized with a salt such as ammonium chloride or ammonium sulfate, and ammonia and acetylene are recovered in a flash tank, where they are separated under vacuum and recycled. The MB crude is freed of acetone as overheads and the MB bottoms are then purified in the second still to hydrogenation, or spec-grade methyl butynol. Heavy ends are processed as bottoms and disposed of.

Figure 1-16B comprises the hydrogenation and dehydration units. MB is hydrogenated at 79 to 91°C in a two-stage reactor at 70 psi using an inhibited palladium catalyst for selective semihydrogenation. The hydrogenation is taken to 99% selectivity and yield. The undistilled methyl butenol (MBe), after filtering from catalyst, is directly used in the dehydration step. The dehydration is carried out at 250 to 300°C using an alumina catalyst that is periodically reactivated with hot air to burn off tars and polymers. The yield of isoprene is almost quantitative under optimum conditions, and the overall yield from acetone averages 90 to 95% of theory. Isoprene is readily purified to high-purity (99.9%) monomer by extractive distillation, which effectively removes C_5 olefins, carbonyl compounds, acetylene, and polar impurities. The extractive distillation also integrates the reworking of reject monomers into the final purification step more effectively than straight distillation.

The SNAM-Progetti-ANIC process can readily yield high-purity (99.9%) isoprene without the expensive purification train required for monomers derived from other processes.

A bonus feature of acetone-acetylene technology is that both methyl butynol (MB) and methyl butenol (MBe) are versatile specialty products. MB is an important building block for the manufacture of vitamins A, E, and K; perfumery compounds, corrosion inhibitors, and chlorinated solvent stabilizers. It is a highly polar compound, soluble in water in all proportions and capable of solubilizing a wide variety of organics, including polymers. MBe has uses as a perfume-flavor and pesticide intermediate. In the years ahead, any significant technology breakthrough in lowering manufacturing costs of acetylene from either coal or perhaps petrochemical sources might make this versatile process also attractive for the manufacture of isoprene in the United States.

Commercial Isoprene Processes The three principal isoprene processes practiced in the United States at present are

Propylene dimerization (Goodyear-Scientific Design) [101]
Dehydrogenation of isoamylene (Shell) [95]
Steam cracking of naphtha (Enjay) [101]

Another important process, practiced in Europe, the Soviet Union and Japan, is the French Institute of Petroleum (IFP) process based on the reaction of isobutylene with formaldehyde (Prins reaction) to yield a dimethyl dioxane derivative that is cracked to isoprene and formaldehyde (Ref. 11, Vol. 2, 102, 103). We shall discuss only the dehydrogenation route to isoprene.

Dehydrogenation of Isoamylene Isoamylene is now considered the most useful C_5 hydrocarbon for dehydrogenation to isoprene. However, it is equally valuable as a component for gasoline, where it was valued in 1967 at 12 ¢/gal [95]. Costs for this material and other hydrocarbons are rising, and how these costs will ultimately effect future process costs for isoprene is speculative. The present value for isoamylene is over 70 ¢/gal.

$$CH_3\text{-}C(CH_3)\text{=}CHCH_3 \text{ or } CH_2\text{=}C(CH_3)CH_2CH_3 \xrightarrow[\text{cat.}]{\Delta} CH_2\text{=}C(CH_3)\text{-}CH\text{=}CH_2 + H_2$$

2-methyl-2-butene (main isomer) or 2-methyl-1-butene (minor isomer)

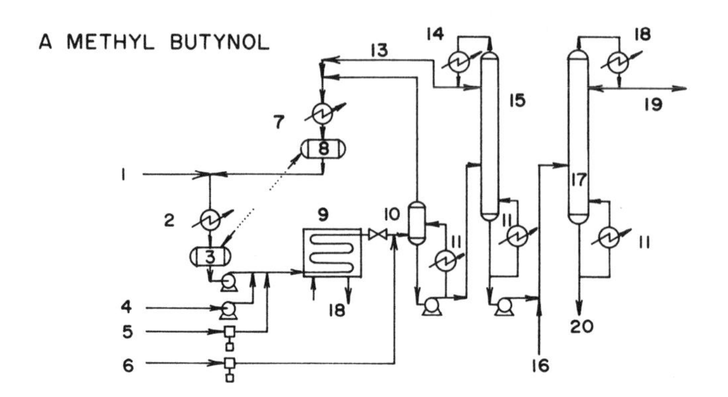
A METHYL BUTYNOL
1
2
3
4
5
6
7
8
9
10
11
13
14
15
16
17
18
19
20

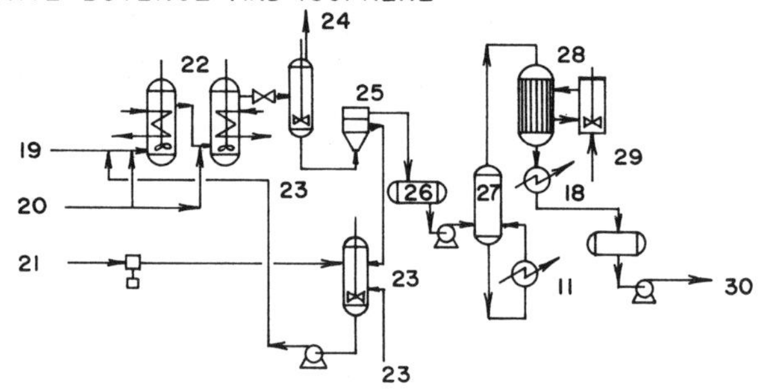
B METHYL BUTENOL AND ISOPRENE
19
20
21
22
23
24
25
26
27
28
29
18
11
30

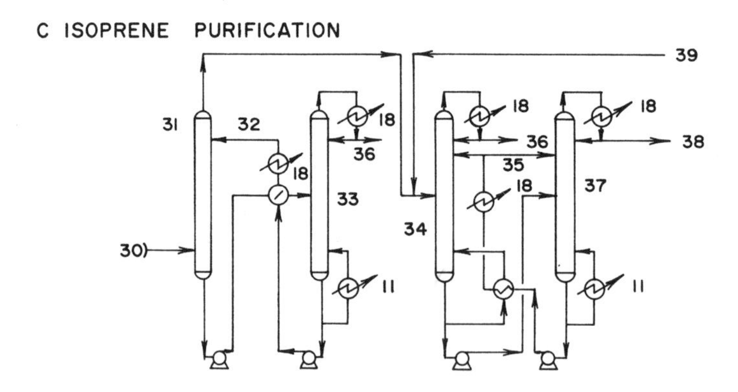
C ISOPRENE PURIFICATION
30
31
32
33
34
35
36
37
38
39
18
11

Besides mixed-branched amylenes, isopentanes can also be used as feed stock [89], although they are less efficient. Isoamylene is extracted from C_5 streams via 65% H_2SO_4 at 50 to 70°F., which is in turn extracted from the acid with a normal paraffin solvent. The distilled isoamylene is dehydrogenated by well-known vapor-phase technology (butylenes, butadiene). Chrome-alumina catalysts are typical for this process. The Shell Chemical Company, utilizing technology pioneered by Houdry [104], has exploited this route commercially and had a rated capacity in 1959 of about 20,000 t/yr at Torrance, California. The isoprene converted to cis-polyisoprene is utilized by U.S. Rubber for truck tires at Los Angeles.

1.6.9 Chlorinated Hydrocarbons

The manufacture of chlorinated solvents from acetylene is well-documented old technology dating back to the preeminence of

---

Figure 1-16 Isoprene from acetylene and acetone (SNAM-Progetti Process). (Reprinted by special permission from *Chemical Engineering*, October 1, 1973. Copyright © 1973, by McGraw-Hill, New York.)

1. Acetylene feed
2. Refrigerant
3. Ammonia condenser
4. Acetone feed
5. KOH catalyst feed
6. Stopper
7. Refrigerant
8. Acetylene condenser
9. Ethynylation reactor
10. Flash tank
11. Steam
12. NH_3 + unreacted C_2H_2
13. Unreacted acetone
14. Refrigerant
15. Acetone recovery column
16. Water
17. Heavy ends column
18. Cooling water
19. Aqueous methyl butynol to hydrogenation
20. Hydrogen feed
21. Inhibitor feed
22. Hydrogenation reactors
23. Fresh catalyst and catalyst feed
24. Unreacted hydrogen
25. Catalyst separator
26. Methyl butenol (MBe) receiver
27. MBe evaporator
28. Dehydration reactor
29. Fuel gas
30. Raw isoprene to purification
31. Washing tower
32. Water to washing tower
33. Water distillation
34. Extractive distillation column
35. Extraction solvent
36. Purge line
37. Isoprene stripper
38. Pure isoprene
39. Reject isoprene from polymerization plant

calcium carbide [105] and represents some of the first commercial chemicals produced from acetylene. This family of products comprises mainly trichloroethylene, perchloroethylene, trichloroethane, tetrachloroethane, pentachloroethane, and hexachloroethane (Ref. 81, pp. 156-170, 188-205). In 1966, in spite of strong competition from cheap starting materials, such as ethylene, methane, propylene, the overall use of both carbide- and petrochemical-based acetylene was still significant. By far, the two most important chlorinated products were trichloro- and perchloroethylenes, which probably account for over 95% of the chlorinated solvents market in the United States. The processes outlined below illustrate the key role of these compounds as both intermediates and end products.

1,1,2,2-Tetrachloroethane and 1,1,3-Trichloroethylene Tetrachloroethane represents a large-volume captive use for the manufacture of trichloroethylene. Acetylene and chlorine are reacted continuously in excess tetrachloroethane (as diluent) in the presence of a catalytic amount of ferric chloride [106] to yield tetrachlorethane, which in turn is dehydrochlorinated with either a lime slurry [81,107], at 100°C or catalytically [108] ($BaCl_2$ on carbon) at 230 to 320°C to yield trichlorethylene.

$$CH{\equiv}CH + 2CL_2 \xrightarrow{FeCl_3} CHCl_2CHCl_2$$

$$\xrightarrow{\Delta\ \frac{1}{2}Ca(OH)_2} \tfrac{1}{2}CaCl_2 + CHCl{=}CCl_2 + H_2O$$

Direct reaction of gaseous acetylene and chlorine without diluents is extremely hazardous and results in violent explosions due to the formation of unstable chloroacetylenes.

Tetrachloroethane, due to its high toxicity, has essentially no industrial use except its conversion to trichlor- and perchlorethylenes. It was at one time used as a fumigant against the white fly in greenhouses and as a potent specialty solvent. Since it is considered one of the most toxic of chlorinated hydrocarbons, it is doubtful if new industrial uses will evolve.

Perchloroethylene (1,1,2,2-Tetrachloroethylene) The original process for this material involved the chlorination of trichloroethylene (derived from C_2H_2 and chlorine) to pentachloroethane [81] and the dehydrohalogenation of the latter to perchloroethylene.

$$CHCl{=}CCl_2 + Cl_2 \longrightarrow CHCl_2CCl_3 \xrightarrow{\Delta} Cl_2C{=}CCl_2 + HCl$$

Elimination of HCl from "pentachlor" can be effected either by lime [81] or catalytically, by vapor-phase technique [109]. No significant commercial application is believed to exist for the pentachloro intermediate.

It was inevitable that the four-step synthesis based on acetylene and chlorine would be contested by more obvious direct routes based on either cheap alkanes or olefins [110,111]. By 1966, only 20% of the perchloroethylene produced in the United States (93 million lb) was based on acetylene.

At present, a variety of C_1-C_3 hydrocarbon (methane, ethane, propane, ethylene, propylene) feedstocks are used via chlorination, oxychlorination, and dehydrochlorination processes. The following is a versatile, multiproduct process based on ethylene:

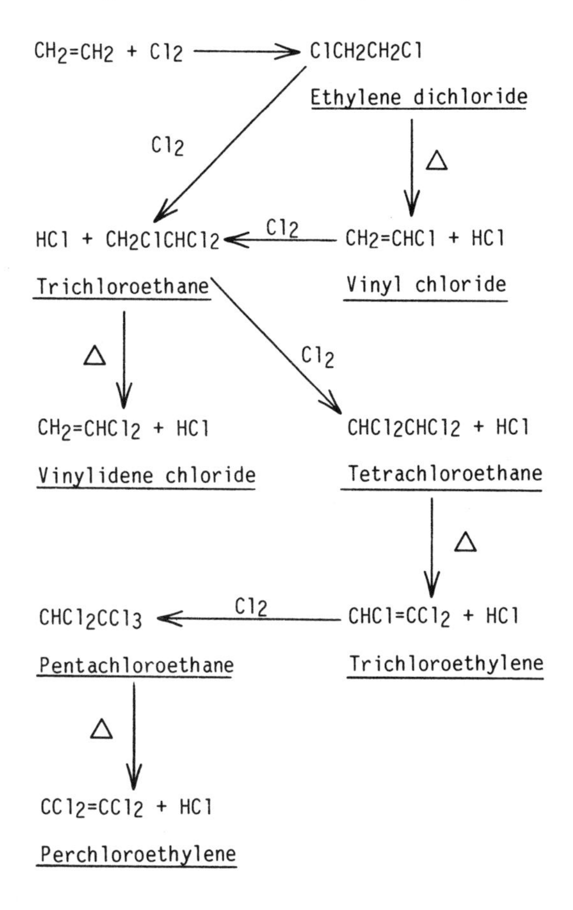

A common product denominator is evident in comparing this process with the acetylene-chlorine route. However, the process is also capable of readily producing large amounts of important products such as vinyl and vinylidene chlorides, making for an integrated, efficient operation. By utilizing new technology in oxychlorination in which by-product HCl can be oxidized to chlorine ($4HCl + O_2 \longrightarrow 2Cl_2 + 2H_2O$), the overall efficiency of the ethylene-based process is increased (cf. Sec. 1.6.10). Also, it is possible to proceed directly from ethylene dichloride to "trichlor" by direct chlorination:

$$ClCH_2CH_2Cl + 2CL_2 \longrightarrow CHCl{=}CCl_2 + 3HCl$$

Dow operates a multiproduct production facility at Freeport, Texas, based on the chlorination of hydrocarbons.

Trichloro- and Perchloroethylenes: Applications, Growth, and Acetylene Consumption Trichloroethylene is used almost exclusively (80%) in the important industrial application of vapor degreasing of metals. Hence, its growth is highly dependent upon growth of the metal treating and fabricating industries and has been estimated at about 6.5%/yr [110]. In 1966, a total of 480 million lb was produced in the United States, 408 million lb of which (85%) were produced from acetylene, the remainder probably being derived from ethylene. During the period 1960-1966, U.S. acetylene consumption for trichlor grew from 69 to 86 million lb. During that time, most of the acetylene used was from carbide (Hooker, Niagara Falls, New York, and Tacoma, Washington; Du Pont, Niagara Falls, New York; Detrex Chemical Industries, Ashtabula, Ohio). Hooker, however, has a new plant at Taft, Louisiana, based on petrochemical acetylene.

Perchloroethylene is used chiefly as a solvent in dry-cleaning applications; in 1967, usage amounted to 435 million lb out of a total production of 535 million lb [112]. Its low toxicity, excellent stability, nonflammability (under normal conditions), ease of recovery, and powerful solvent action make it eminently useful in the dry-cleaning trade. These premium properties have enabled it to gradually replace Stoddard solvent (petroleum distillate) in many dry-cleaning applications, in spite of its higher cost. It is extensively employed in coin-operated dry-cleaning systems and, as of 1967, had approximately 55% of the total dry-cleaning market [113].

1.6.10 Vinyl Chloride

Vinyl chloride (VC) production is today characterized by huge production facilities in which only chemical giants such as B. F. Goodrich, Union Carbide, Dow, and Shell can economically operate. The vinyl chloride-polyvinyl chloride business is a low-profit operation where competition in the market place is intense. The marginal producer cannot exist in this competitive environment. Large PVC producers are often back-integrated to VC production and forward-integrated to consumer markets to aid overall profits. Acetylene has now been largely displaced in the United States as a VC feedstock in favor of ethylene. The trend has also been in this direction both in Europe and in Japan.

Acetylene-Hydrogen Chloride (Hydrochlorination) Process The only United States producer currently using the acetylene-based route for the production of vinyl chloride is Monochem, at their Geismar, Louisiana site, where acetylene is also used to manufacture vinyl acetate as part of a large, balanced petrochemical complex (Sec. 1.6.11). The hydrochlorination process, however, has had a long history of efficient, trouble-free performance, and in the period 1950-1965 accounted for a major portion of the acetylene used for commodity chemicals manufacture.

Figure 1-17 is a process flow diagram for vinyl chloride production. The heart of the process is the fixed-bed (plug-flow) reaction system [114-116] (tube and shell heat exchangers) packed with mercuric chloride (10%) on a granular carbon support [116]. Acetylene compressed (30 psia), dried, and freed of impurities via a purification train is mixed with excess dry HCl in 1:1.1 mole ratio in the reactor. Unreacted acetylene and entrained vinyl chloride are recycled to the reactor, and both reaction streams are processed to monomer grade vinyl chloride by a combination of scrubbing (water, caustic), absorption, stripping, and pressure distillation.

The hydrochlorination of acetylene is characterized by high (98-99%) conversion of both reactants and high selectivity (98%) to vinyl chloride (Ref. 11, Vol. 2, pp. 72-89, Ref. 114). The process can also be run effectively at lower production rates where ethylene-based technology is less economical. The reaction is exothermic ($\sim$25 kcal/g mol) and heat transfer fluid circulated through the shells of the reactors controls temperature in the catalyst bed at about 90 to 140°C. With fresh, active catalyst (10% mercuric chloride on carbon) hydrochlorination takes place readily at 90°C, and only when activity decreases is the bed temperature gradually raised. Catalyst life averages more than 6

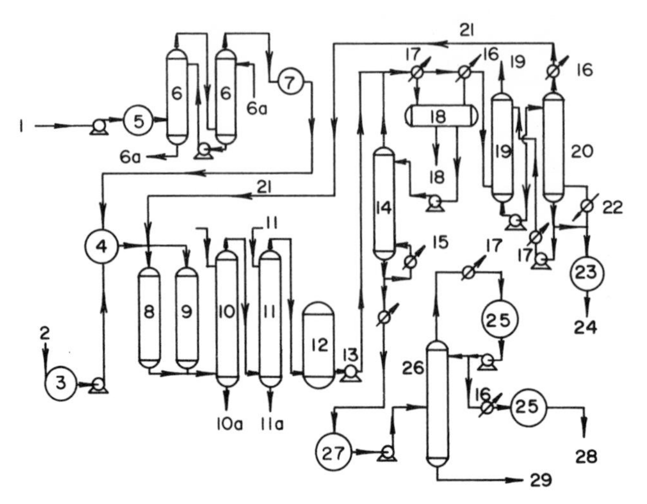

1. Acetylene feed
2. HCl feed
3. Purifying section
4. Mixing tank
5. Entrainment separator
6. Drying columns

6a. H_2SO_4 feed to 75% H_2SO_4

7. Activated carbon filter
8. Recycle reactor
9. Primary reactor
10. Water scrubber

10a. 10% HCl from scrubber

11. Caustic feed-caustic scrubber

11a. Spent caustic from 11.

12. H_2O knock-out tank
13. Reciprocating compressor
14. Stripping column
15. Hot water
16. Cold brine
17. Cold water
18. Decanter and water effluent
19. Absorber and light ends
20. Stripper (C_2H_2, VCM, heavy ends)
21. Untreated C_2H_2 and VCM
22. Steam
23. Storage tank
24. Heavy ends
25. Hold tanks (VCM)
26. Distillation column
27. Storage tank
28. Purified vinyl chloride
29. Heavy ends

Figure 1-17 Acetylene-hydrogen chloride route to vinyl chloride. (Reprinted by special permission from *Chemical Engineering*, March 27, 1967. Copyright © 1967, by McGraw-Hill, New York.)

months, since the catalyst is not readily poisoned except for sulfur. By-products are formed in trace amounts and include dichloro- and trichloroethylenes, besides acetaldehyde.

It is unlikely that new acetylene-based VCM plants will be built in the near future in the United States due to the high cost of petrochemical acetylene versus ethylene and ethane. However, the combination of cheaper oil stocks and the desire to forward integrate into petrochemicals could make the more simple and trouble-free acetylene route attractive to Arab nations and Iran. In these countries enormous amounts of low-molecular-weight hydrocarbons are burned (flared), which could otherwise be recovered and used for acetylene production. In the United States the principal hope for an acetylene-based vinyl chloride process lies ultimately in cheaper acetylene from coal-based technology (cf sections 1.2, 1.3.2.5, and 1.7.2 of this chapter).

Balanced Ethylene-Acetylene Vinyl Chloride (VC) Process The balanced process concept was already established commercial practice in the United States in the 1960s when the cheaper feedstock ethylene was used with acetylene to lower VC process costs. By utilizing or "balancing" the hydrogen chloride formed by the dehydrohalogenation of 1,2-dichloroethane (DCE) to make additional VC by reaction with acetylene, a hybrid process resulted, more economical than either route alone:

(a) $CH_2{=}CH_2 + Cl_2 \longrightarrow ClCH_2CH_2Cl$ (DCE)

(b) $DCE \xrightarrow{\Delta} CH_2{=}CHCl + HCl$

(c) $CH{\equiv}CH + HCl \longrightarrow CH_2{=}CHCl$ (VC)

Overall Process (a) + (b) + (c):

$$CH_2{=}CH_2 + CH{\equiv}CH + Cl_2 \longrightarrow 2CH_2{=}CHCl$$

In recent years this technology [115] has become less important as acetylene production facilities from either calcium carbide or petrochemical sources have been closed down. In its place has been substituted the oxychlorination concept, in which HCl is oxidized to chlorine and used to chlorinate ethylene to form DCE [105, 117-119]. The balanced ethylene-acetylene route has been modified to realize more favorable economics.

In Japan, Kureha Chemical Industry Co. Ltd. has optimized this technology [119,120] by the use of sophisticated in-line computer control. This method eliminates the more costly separation of ethylene and acetylene from the pyrolysis stream, while still utilizing the advantages of the balanced system. Since the latter system has been gradually replaced by oxychlorination technology, this approach at present is of questionable merit in the United States.

However, in the mid-1980s, as greater amounts of crude oil are cracked to form larger amounts of by-product acetylene (see Sec. 1.1 and 1.3.2.4 of this chapter), this process could become economically attractive. A flow diagram of the Kureha process is shown in Fig. 1-18.

Balanced Ethylene Oxychlorination Vinyl Chloride (VC) Process

In practically all variations of this technology, by-product HCl resulting from the cracking of dichloroethane (DCE) is oxidized to chlorine by the use of a mixed salt catalyst, comprised mainly of copper chlorides deposited on porous supports such as alumina, silica, or silica-aluminates [115]. Such catalysts are commonly known as *Deacon catalysts* when applied to the oxidation of HCl. In the oxychlorination process, recycled (recovered) HCl, ethylene, and oxygen are reacted simultaneously in a continuous system to form DCE. In a separate coreaction system, ethylene is chlorinated to DCE, which is then pyrolized to VC and HCl. The steps involved in the overall process are:

(a) $2HCl + \frac{1}{2}O_2 \longrightarrow Cl_2 + H_2O$ (Deacon reaction)

(b) $CH_2{=}CH_2 + 2\ HCl + \frac{1}{2}O_2 \longrightarrow \underset{\text{DCE}}{ClCH_2CH_2Cl} + H_2O$ (oxychlorination)

(c) $CH_2{=}CH_2 + Cl_2 \longrightarrow DCE$

(d) $DCE \longrightarrow \underset{\text{VC}}{CH_2{=}CHCl} + HCl$ [used in (b)]

Overall reaction (b) + (c) + (d):

$$2CH_2{=}CH_2 + Cl_2 + \tfrac{1}{2}O_2 \longrightarrow 2VC + H_2O$$

This technology is now being practiced worldwide (Ref. 11, vol. 2, pp. 72-84).

The chlorination of ethylene to dichloroethane (DCE) proceeds in close to quantitative yield. The purification of DCE is a relatively simple step, since no significant amounts of by-products are formed.

Hydrogen chloride fractionated from vinyl chloride is recycled to the oxychlorination reactor, where it reacts with ethylene and oxygen to form DCE, which in turn is purified by distillation in the DCE purification unit. DCE in the cracking furnace is converted to VC and HCl in 99% yield.

Chlorination of Ethylene to 1,2-Dichloroethane (DCE) The chlorination step is common to both the balanced and oxychlorination processes for the manufacture of vinyl chloride. The reaction is carried out continuously using either liquid- [119,121] or vapor-phase [122,123] systems. Both types of processes are generally run under pressure (4-10 atm) and at temperatures of 50 to 140°C. Yields of vinyl chloride by both routes are excellent, with 99% of the chlorine utilized for VC production. Usage of ethylene is 95- to 98% of theory, with similar selectivity to VC. The vapor-phase process is run with a higher excess of ethylene. In both processes by-product formation (C_2HCl_3 and C_2Cl_4) is quite low.

The liquid-phase process, particularly the reactor section, is simple to design, erect, and operate compared with the vapor-phase system. Dichloroethane is used as the reaction solvent and as a heat transfer media to control the resulting exothermic reaction between ethylene and chlorine. Reaction temperatures are easy to control in the 50 to 70°C range, and larger excesses of ethylene or an inert gas diluent are not required. The resultant heat of reaction from the chlorination can be conveniently used to preheat the reactants and to also vaporize DCE formed, for the subsequent purification step.

Pyrolysis of 1,2-Dichloroethane (DCE) to Vinyl Chloride The Goodrich high-pressure pyrolysis process for the production of vinyl chloride is a standard for the PVC industry and has helped to make the B. F. Goodrich Company one of the world's largest producers of PVC. Although over the years numerous patents (Ref. 11, Vol. 2, pp. 72-84;115) have been issued describing catalytic or inert solid pyrolysis of DCE to VC, the simple noncatalytic, high-pressure method is still preferred. Goodrich alone produces over 1 billion lb of vinyl chloride annually at their Calvert City plant using this process.

Figure 1-19 is a flow diagram of the pyrolysis process, containing the cracking furnace, cooling (quench) and distillation assemblies, and recycle system for DCE [124]. The multi-tube furnace is operated in the last three sections at 470 to 540°C and 24 to 25 atm, and the conversion of DCE per pass to vinyl chloride averages 50 to 70%, at a residence time of 9 to 20 sec. Overall yields of VC are over 99% based on DCE. By-products are quite low and include acetylene, benzene, methyl chloride, tars, and

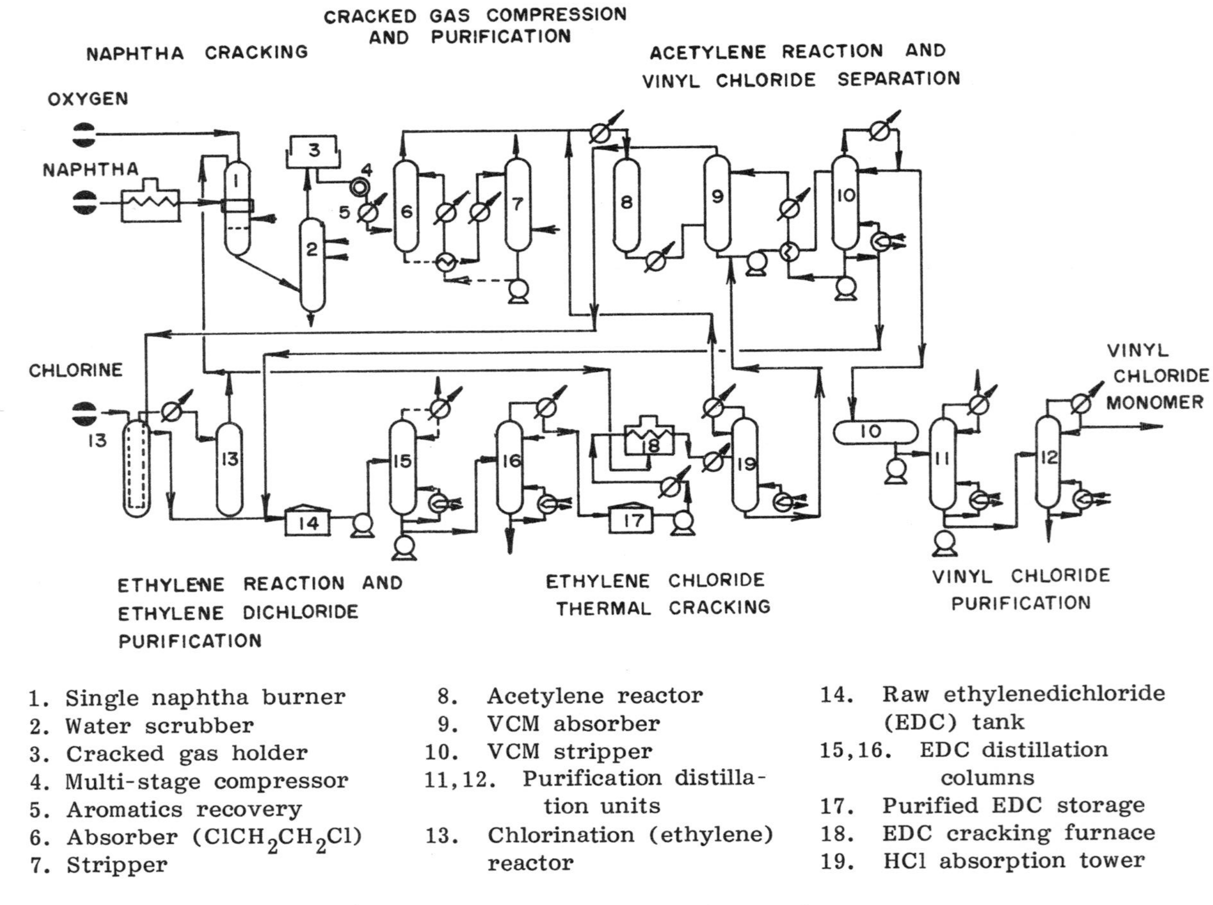

Figure 1-18 Kureha process for vinyl chloride-naphtha based.

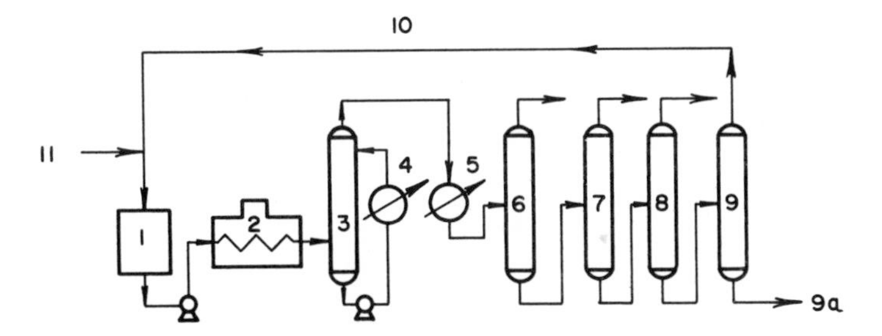

1. Storage tank
2. Pyrolysis furnace
3. Quench column
4,5. Heat exchangers
6. HCl stripping column
7. Vinyl chloride distillation
8. Light ends column
9. Dichloroethane (DCE) column and heavy ends (9a)
10. DCE recycle
11. DCE feed to 1.

Figure 1-19 High pressure pyrolysis of dichloroethane (DCE).

coke. The latter two by-products build up slowly in the pyrolysis tubes and are removed by burning them out with air.

The gases emerging from the furnace are cooled with crude DCE (quench column) to their dew point temperature before being separated in the distillation train. The first distillation still recovers HCl gas for oxidation to chlorine (oxychlorination), while the second distillation column produces monomer-grade vinyl chloride. Low boiling (light ends) impurities are removed in the third column, before DCE is separated for recycling from higher boiling by-products (heavy ends) in the fourth still.

1.6.11 Vinyl Acetate (VAM)

In the United States vinyl acetate production based on acetylene is now less than 15% of total production. During the 1960s it was second to vinyl chloride as a user of acetylene. It is possible that in Europe and Japan acetylene-based VAM is proportionately higher. In the 1970s the vapor-phase ethylene-based technology for VAM is dominant in the United States and probably the rest of the world [125].

Vinyl Acetate from Acetylene The catalytic addition of acetic acid to acetylene to form vinyl acetate was practiced as both liquid- and vapor-phase processes [126]. The vapor-phase method, being more efficient, eventually replaced the liquid-phase process. The catalyst of choice has continued to be zinc acetate (20-30% by weight) deposited on granular carbon. A typical flow diagram for

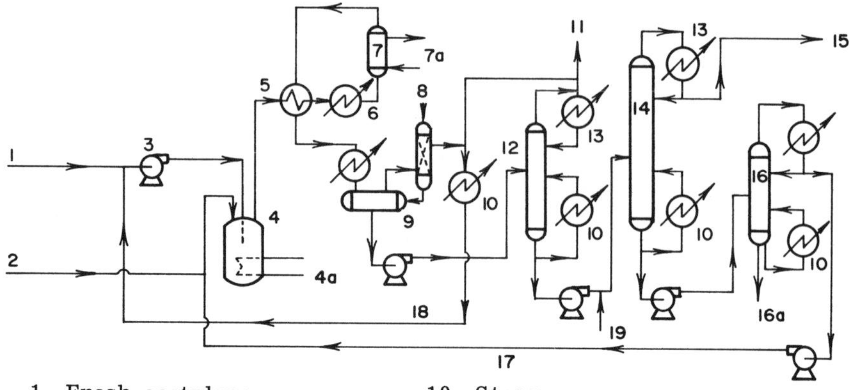

1. Fresh acetylene
2. Fresh acetic acid
3. Blower
4. Charge vaporizer
4a. Steam-heating coils
5. Reactor charge
6. Steam
7. Reactor
7a. Reactor dowtherm
8. Acetic acid
9. Crude vinyl acetate surge
10. Steam
11. Purge
12. Light ends column
13. Cold water
14. Vinyl acetate column
15. Finished vinyl acetate
16. Heavy ends column
16a. Heavy ends
17. Recycle acetic acid
18. Recycle acetylene
19. Inhibitor feed

Figure 1-20 Acetylene-acetic acid process for vinyl acetate.

this process is described in Fig. 1-20. In the mid-1960s this technology was practiced extensively at production sites at Calvert City (Air Reduction Company); Niagara Falls (Du Pont and Union Carbide); Shawinigan Falls, Quebec (Shawinigan Chemicals, Ltd.) and Texas City, Texas (Union Carbide). These plants have since been closed down in favor of ethylene-based technology.

Acetylene for this process [127] must be free of phosphine and sulfur impurities to avoid catalyst poisoning. A typical treatment for calcium carbide acetylene is to wash it with a chromic-sulfuric acid mixture to remove these impurities. Acetic acid (both fresh and recycled) is vaporized in a stream of acetylene in the charge vaporizer at about 5 psig and 71 to 82°C. The reaction charge (acetic acid-acetylene) is preheated to about 177°C with a fresh catalyst charge, but near the end of the catalyst's life (2-3 months) temperatures as high as 210°C are reached. The temperature range (reactor) averages from 180 to 210°C.

The process [127,128] is run with a molar ratio of C_2H_2 to acetic acid of 4-5:1, at 2 to 3 psig, and a space velocity of 300-400 hr^{-1}.

The temperature rise (exotherm) in the bed averages 5 to 10°, and the heat of reaction is removed with heat transfer fluid from the shell of the reactor. The overall process yields are 92 to 98% on actylene and 95 to 99% on acetic acid. The vinylation reaction is clean, with no tar formation, and the principal by-products are acetaldehyde and ethylidene diacetate.

The reactor section [127,128] comprises a typical shell and tube reactor, with the tubes about 2 in. in diameter, 12 ft long, and packed with 20 to 30% zinc acetate on granular carbon catalyst. The reaction is characterized by a "hot spot" zone that migrates as catalyst activity decreases. In about 2-3 months a fresh charge is usually required to maintain desired production rates.

The effluent gas from the reactor, consisting mainly of crude vinyl acetate, is first cooled by exchange with feed vapors and then condensed with coolant. Uncondensables, consisting mainly of acetylene and possibly acetaldehyde, pass overhead and are recycled to the reactor. The condensed crude, mainly vinyl acetate and acetic acid with minor amounts of acetaldehyde, acetone and ethylidene diacetate, is processed to "spec" grade vinyl acetate monomer (VAM) by the three distillation columns shown in Fig. 1-20. The process handles acetylene at low pressures (5 psig) diluted with acetic acid and, hence, is not considered hazardous. It has a long history of safe operation. However, the only producer of VAM by the acetylene route in the United States at present is Monochem (Borden-Uniroyal). Whether this excellent process will be used again depends on the future costs of acetylene, most likely coal based, compared with ethylene or ethane.

Vinyl Acetate from Ethylene Arganbright [128] has reviewed important vinyl acetate processes, including the earlier ICI process, which involved the liquid-phase reaction of ethylene, oxygen, and acetic acid using a palladous chloride catalyst. Since then, the liquid-phase process has been abandoned due to severe corrosion problems and production difficulties. The preferred technology [125] now practiced worldwide, is to react ethylene, oxygen, and acetic acid in the vapor phase over a supported palladium metal catalyst.

Figure 1-21 is a flow diagram of the vapor-phase process [125].

The vapor-phase technology [129,130] is characterized by high conversions and selectivity to vinyl acetate, with essentially no acetaldehyde by-product formation or significant corrosion of equipment. Yields to VAM based on ethylene average 91 to 94%, with acetaldehyde formation at 1% or less. The catalyst used is palladium or a mixture of noble metals and salts (also alloys) deposited on a suitable support. Catalyst concentration varies from

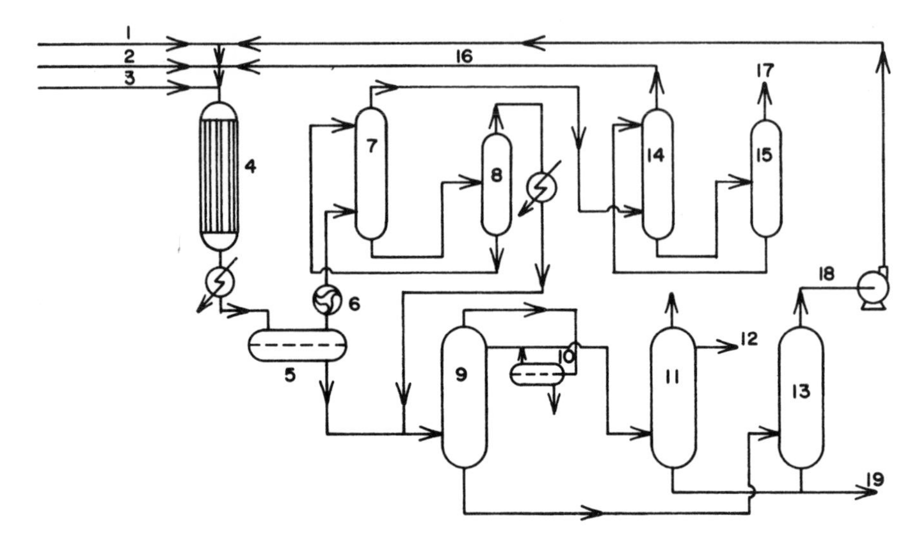

1. Acetic acid feed
2. Ethylene feed
3. Oxygen feed
4. Reactor-shell and tube type
5. Reactor condensate
6. Booster
7. Glycol wash
8. Desorber
9. Azeotropic distillation
10. Water-organic phase separator
11. Vinyl acetate distillation
12. Vinyl acetate product
13. Acetic acid distillation
14. Hot carbonate wash
15. Desorber
16. Recycle gas
17. CO_2 off gas
18. Acetic acid recycle
19. Heavy ends impurities

Figure 1-21 Ethylene-acetic acid-oxygen vapor-phase process for production of vinyl acetate.

0.1 to 2.0% palladium. Other noble metal combinations can be used in amounts of 10 to 70% of the palladium catalyst.

The process involves vaporizing acetic acid at 120°C in a stream of ethylene (5-10 atm gauge) by usual techniques. Oxygen is fed to the vaporized feed just before the reactor and below the explosive limits (5% oxygen) of the mixture. The reactants are passed over the catalyst bed at 175 to 200°C, and the heat of reaction is used for various other process needs. The conversions per pass of reactants average 60 to 90% (oxygen), 15 to 30%

(acetic acid), and 10 to 15% (ethylene) at space velocities of 200 g vinyl acetate per liter of catalyst per hour [126,128,129]. The oxidation of ethylene to carbon dioxide averages about 10% of theory.

Isolation of vinyl acetate is effected by cooling reactant gases, followed by separation of liquid from gaseous products and recycling the latter (Fig. 1-21). Prior to recycling, the gases are scrubbed with propylene glycol to remove vinyl acetate, then extracted with hot sodium carbonate solution to remove by-product CO_2. Vinyl acetate desorbed from the glycol wash is combined with crude vinyl acetate from the reactor, then purified through azeotropic distillation (water removal), followed by successive distillations to separate light ends, vinyl acetate, acetic acid, and heavy ends, in which three distillation columns are used (Fig. 1-21).

1.7 OVERVIEW: COMMODITY CHEMICALS AND ACETYLENE

The large decrease in acetylene consumption for the production of large-volume chemicals is shown in Table 1-5 for the years 1970, 1974, and 1979 (53,9]. The period 1970-1974 has been particularly significant.

The period 1970-1974 witnessed the closing down of a number of carbide and petrochemical facilities (see Sec. 1.5.3) and the substitution of processes using cheaper feedstocks, mainly ethylene. Propylene is now the predominant feedstock for acrylates. BASF (Freeport, Texas) [132] is currently the only producer using the acetylene carbonylation process, but in 1982 it plans to switch all its production to the propylene-based technology. As summarized in Table 1-5, the principal growth area for acetylene in the 1980s will probably be acetylenic chemicals and derivatives. The overall growth rate of acetylene is projected to be near zero up to 1982 [132].

Although acetylene has been largely replaced in the manufacture of vinyl acetate and vinyl chloride, it is still used by Borden and National Starch as part of a balanced petrochemical complex. Hooker also closed down its chlorinated solvents plant early in 1978, which used about 10 million lb of acetylene annually, for the production of trichloroethylene and perchloroethylene.

Over a 10-year period (1965-1974) total acetylene consumption decreased from a high point of 1,230 to 507 million lb, averaging about 9.4% annually [52,9]. Most of the loss, however, took place between 1970 and 1974, as shown below. The large price differential between acetylene and ethylene was the prime reason for this decline, shown in the tabulation. Chemical manufacturing accounted for most of this decline, since industrial uses such as metal fabrication, welding, and cutting are relatively constant and average about 100 million lb annually. The period 1967-1974 saw the loss of about 1 billion lb of acetylene consumption, or the equivalent of about 3 billion lb of plant capacity, for the production of commodity chemicals. During this period a total of 23 plants that manufactured such chemicals as vinyl chloride, vinyl acetate, chloroprene (neoprene), chlorinated solvents, and acrylonitrile were closed down. The majority (16) of these plants manufactured vinyl chloride and vinyl acetate.

		Average Bulk Price (¢/lb)	
Year	Acetylene Used (MM lb)	Acetylene	Ethylene
1965	1230	7-12	3-4
1966	1226		
1967	1065		
1968	1120		
1969	1195		
1970	1023		
1971	852	7-15	3-5
1972	790		
1973	571		
1974	507		
1976	490	14-20	8-9
1977	310	25-30	9
1979	269	30-50	14
1983 (est.)	269	55-75	23-29

The above acetylene usage from 1965 to 1976 includes about 100 million lb/yr of acetylene for industrial uses (cutting and welding), while the 1977-1983 figures reflect only chemical use.

Table 1-5 USA Chemical and Industrial Acetylene Usage

	Millions of Pounds		
Product	1970	1974	1979
Acrylates	70	105	45
Acrylonitrile	42	0	0
Chloroprene (neoprene)	242	0	0
Chlorinated solvents	91	15	0
Vinyl chloride	268	125	100
Vinyl acetate	158	73	37
Acetylenic chemicals[a]	41	69	73
Other[b]	10	15	14
Total Chemical Use	922	402	269
Industrial use (welding)[c]	101	105	122
Total uses	1023	507	391

[a] Acetylenic chemicals: mainly butane-1,4-diol, plus tetrahydrofuran (THF), acetylenic alcohols and glycols, and miscellaneous specialties.
[b] Other: mainly vinyl fluoride and acetylene black.
[c] Industrial use: metal cutting, fabrication, and welding.

1.7.1 Acetylenic Chemicals and Derivatives

The greatest growth area for acetylene is now in Reppe-type chemicals such as butyndiol, butanediol, tetrahydrofuran (THF), N-methyl-2-pyrrolidone, vinylpyrrolidone, and polyvinylpyrrolidone, besides a variety of smaller volume acetylenic alcohols and glycols [131]. The technology and applications for large-volume acetylenic products and derivatives are covered in Chap. 2, together with specialty acetylenic alcohols and glycols and their derivatives.

Total acetylene usage for acetylenic chemicals and derivatives in 1982 is projected to 106 to 121 million lb, which represents 8 to 11% annual growth from 1977 to 1982. The consumption of acetylene for these chemicals is compared in the following table:

Product	C_2H_2 (MM lb)	
	1977	1982 (est.)
Butane-1,4-diol	66	110-115
Vinyl ethers	5	5
Acetylenic alcohols and derivatives	1	2
Vinyl fluoride	4-5	>5

Gulf will manufacture acetylene black [131] in 1979 at its Cedar Bayou, Texas site using by-product acetylene from its olefins production facility. The capacity of the new plant is rated at 15 to 20 million lb/yr. A similar plant at Ponce, Puerto Rico, owned by Union Carbide, is rated at 8 million lb. Continued growth over the years is projected for this unique, conducting specie of carbon.

Vinyl fluoride is being produced at the Du Pont plant at Louisville using calcium carbide acetylene from the nearby AIRCO-BOC generating plant. The acetylene consumption in 1977 was estimated at 4 to 5 million lb, which yielded about 7 million lb of vinyl fluoride or polyvinyl fluoride (PVF). The principal use for PVF is in the manufacture of Tedlar film, sheetings and coatings, primarily for exterior uses. Growth in this product, particularly in exterior home siding, is expected to continue over the years.

Du Pont utilizes acetylene from the Tenneco production site (capacity 100 million lb) at Houston and for the production of tetrahydrofuran (THF) and butane-1,4-diol. An important growing use for butanediol is the manufacture of the engineering plastic polybutyleneterphthalate.

1.7.2 Coal: Springboard to the Future

During the period 1979-1980 numerous articles appeared in the chemical press heralding the onset of the age of coal, and the resultant energy and chemicals that could be derived from it. Most of these articles were optimistic [133,134] but some were shaded with caution and pessimism [135-137]. Some key problems that have been cited as related to the large-scale use of coal are the huge cost to the United States, increased air pollution, greenhouse effect (atmospheric CO_2 buildup), choice of technologies to be scaled up, political policies, and economic viability versus current and future oil production. Conversion processes now being extensively studied include conventional coal gasification [134], underground gasification [138], direct and indirect coal liquification [136,140], fluidized-bed combustion [139] and solvent refined coal (SRC) [133].

In South Africa a lack of oil and natural gas, together with political and geographical isolation, has necessitated this country's utilization of its rich coal resources. Since the mid-1950s, indirect coal liquification (a blending of gasification and Fischer-Tropsch technology) has been practiced commercially in South Africa (see Sec. 1.1) to supply fuel oil and various hydrocarbons for its industrial needs. The earlier production facility, known as Sasol I, is now being augmented by a huge, new coal-liquification complex known as Sasol II and III, the total cost of which has been estimated at $7 billion [140]. Initial production of synfuel from Sasol II is scheduled to begin during 1980, and the entire complex should be complete by 1983. The facility, when completely operational, will supply most of South Africa's energy needs in the form of gasoline, jet, diesel, and other related fuels.

Since July 1979 the United States has made a commitment to a multibillion dollar program to free itself from dependence on foreign oil, partly by the utilization of its abundant coal reserves [135]. Initial large-scale pilot plants have demonstrated that the direct liquification of coal using hydrogen and a solvent at high temperatures and pressures may be preferable to the South African technologies [136]. The well-practiced indirect liquification method remains, however, as back-up technology.

A further step to coal utilization in the United States has been taken by Tennessee Eastman, an operating division of Eastman Kodak Company [141], whereby coal will be used to produce a variety of chemicals as shown below. The desired end product is acetic anhydride, used by Eastman Kodak to manufacture cellulose acetate for photographic film base, fibers, and plastics.

(a) Carbon monoxide and hydrogen (synthesis gas):

$$3C + H_2O + O_2 \text{ (heat source)} \longrightarrow 3CO + H_2 + CO_2 \text{ (minor)}$$

(b) Sulfuric acid (recovered sulfur impurity in coal):

$$2S + 3O_2 \longrightarrow 2SO_3 + 2H_2O \longrightarrow 2H_2SO_4$$

(c) Methanol:

$$CO + 2H_2 \longrightarrow CH_3OH$$

(d) Methyl acetate:

$$CH_3OH + CH_3CO_2H \longrightarrow CH_3CO_2CH_3 + H_2O$$

(e) Acetic anhydride:

$$CH_3CO_2CH_3 + CO \longrightarrow CH_3CO\text{-}O\text{-}COCH_3$$

The present usage of acetic anhydride by Eastman Kodak is over 1 billion lb annually. Up to now this compound was exclusively ethylene based. The new coal-based process is scheduled to go on stream in 1983 at Kingsport, Tennessee, but its capacity has not been disclosed at present. It has been estimated that to reduce ethylene consumption by 200 million lb annually, the new acetic anhydride complex would have to have a capacity of about 730 million lb/yr. The present ethylene capacity of Eastman at Longview, Texas is about 1.3 billion lb, and present requirements in terms of acetaldehyde, acetic acid, and acetic anhydride are straining this capacity. The investment cost [10] for new ethylene capacity in the 1980s is considered to be unfavorable, which accounts for Eastman Kodak's switch to coal-based technology.

The production of acetylene from coal, in a broader definition, can be considered a special form of gasification, although more energy intensive. The early manufacture of commodity chemicals was coal based, since acetylene, the raw material of choice, was made from coke and limestone. In the past, the high energy requirements to produce acetylene by either the carbide or AVCO processes (see Sec. 1.2 and 1.3.2) were economic barriers when compared with the cheaper thermal cracking or partial oxidation processes used for olefins and petrochemical acetylene. This situation is changing rapidly since petroleum-based feedstock and natural gas have escalated markedly in price in recent years. Besides coal, organic waste material, (wood, plants, biomass) readily converted to carbon are possible raw materials for future acetylene production.

The continuing national commitment to nuclear energy could mean that this energy source may ultimately become cheaper than energy derived from petroleum. Liquid-metal (Na) fast breeder reactors can convert abundant uranium 238 to fissionable plutonium 239, and can provide an energy source theoretically secure for hundreds of thousands of years, since more plutonium is regenerated than is used up [142,143,144]. Fusion energy, the other side of the nuclear coin, has great future potential for producing abundant and cheaper energy via the fusion of isotopes. The most promising system at present is based on hydrogen isotopes (deuterium-tritium) and is capable of releasing enormous amounts of energy via fusion to helium, in an earthly model of the sun.

The energy available from these nuclear processes would be ideal for the high-temperature conditions needed to produce acetylene by either the AVCO or calcium carbide technologies:

AVCO

$$2C + H_2 \longrightarrow H\text{-}C\equiv C\text{-}H \quad \text{(plasma)}$$

Calcium carbide:

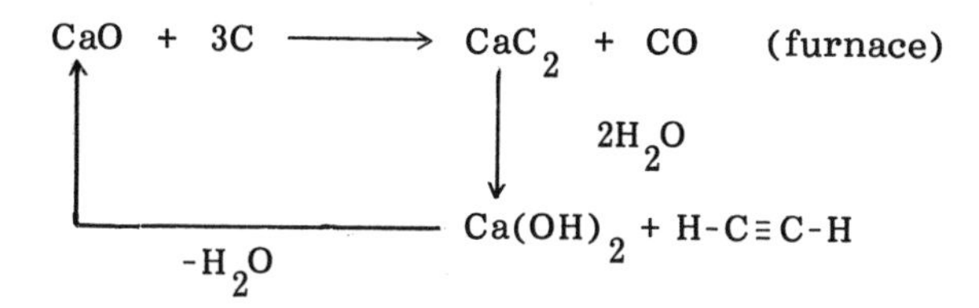

Based on the above considerations it would appear that the longer range potential of acetylene as a chemical raw material for the production of commodity chemicals is attractive. Acetylene, however, has both present and future value for the production of specialty products and new, unique chemicals with emerging-growth potential. These products and uses are described in Chap. 2.

REFERENCES

1. *Time*, OPEC's painful squeeze, July 9, 1979, pp. 12-13; Another oil price stunner, December 24, 1979, pp. 60-61.
2. G. Haber, The fuel of the twenty first century will be coal, *The Sciences*, March 1974, p. 21.
3. *Chem. Week*, Chemicals from coal: Best bet in the energy crisis, June 12, 1974, p. 21.
4. *Chem. Eng. News*, Natural gas—The next process industry looms, August 30, 1971, p. 8.
5. D. L. Burke, CW report, *Chem. Week*, September 11, 1974, p. 38.

6. R. E. Kirk and D. F. Othmer, *Encyclopedia of Chemical Technology, Vol. 6,* 3rd ed., Interscience, New York, 1979, pp. 224-298.
7. R. E. Kirk and D. F. Othmer, *Encyclopedia of Chemical Technology, Vol. 4,* 2nd ed., Interscience, New York, 1964, pp. 278-280, 446-456.
8. S. C. Stinson, Ethylene technology moves to liquid feeds, *Chem. Eng. News,* May 28, 1979, pp. 32-36.
9. T. H. McGreevey, *Chemical Economics Handbook, Acetylene,* 605.5021, *A-V,* August 1975, Stanford Research Institute (SRI), Menlo Park, Calif.
10. *Chem. Eng. News,* Uncertainties plague ethylene industry, May 28, 1979, pp. 11-12.
11. S. A. Miller, *Acetylene, its Properties, Manufacture and Uses,* (in 2 Vols.), Academic, New York, 1965, 1966, *Vol. 1,* pp. 165-282, 283-370, 331-342, 371-375, 384-391, 393, 405-406, 456-469, 484-542; *Vol. 2,* pp. 76, 72-84, 156-183, 134-147.
12. V. J. Clancy, Liquid and solid acetylene-a review of published information, Explosives Research and Development Survey, British Government Publication, England, January 5, 1951.
13. R. E. Kirk and D. F. Othmer, *Encyclopedia of Chemical Technology, Vol. 1,* 2nd ed., Interscience, New York, 1963, pp. 181-193, 338-350.
14. S. A. Miller, *Chem. Ind.,* January 5, 1963, pp. 4-16.
15. Underwriters Laboratories, Inc., *Standards for Acetylene Generators,* No. 297, *Portable Medium Pressure,* May 1973; No. 408, *Stationary Medium Pressure,* May 1973.
16. E. Schallus and A. Goetz, German Patent 1,012,889, August 1, 1957.
17. A. V. Grosse, *U.S. Gov. Reports 37*(12), 61, June 20, 1962.
18. R. E. Kirk and D. F. Othmer, *Encyclopedia of Chemical Technology, Vol. 1,* 3rd ed., Interscience, New York, 1978, pp. 84-89, 195-200, 211-237, 227-231, 231-232, 414-426.
19. I. G. Farben-Electric Arc Patents: DRP, 615,835 (1935); 630,825 (1936); 637,187 (1936); 639,662 (1936); 641,091 (1937); 652,348 (1937); BIOS 1048.
20. H. Gladisch, *Hydroc. Proc. Pet. Ref., 41*(6), 159 (1962).
21. *Chem. Eng. News,* Du Pont electric arc process, *39*(10), 23 (1961); *Chem. Week 94*(3), 64 (1964).
22. G. Doukas, U.S. Patent 3,073,769 (January 15, 1963).
23. M. T. Cichelli and W. Schotte, British Patent 938,823 (January 15, 1963).

24. BASF (Sachsse) Process: BIOS 877 and 1048; FIAT 426 and 988 Reports, U.S. Department of Commerce; H. Sachsse, *Chem. Ing. Tech. 21,* 129 (1954); *26,* 245 (1954).
25. H. Sachsse, T. Kosbahn, and E. Lehrer, U.S. Patents 2,715,648 (1955); 2,719,184 (1955).
26. F. A. Braconier, U.S. Patent 3,019,271 (May 18, 1959).
27. G. Fauser, Italian Patents 578,006 (1958); 624,115 (1961); British Patent 846,368 (1960); *Chem. Week, 87*(23), 60 (1960).
28. F. Fauser, *Chim. Ind. (Milan) 42*(2), 150 (1960).
29. M. S. P. Bogart, U.S. Patent 2,796,951 (June 24, 1957).
30. G. H. Bixler and C. W. Coberly, *Ind. Eng. Chem. 45*(12), 2596 (1953).
31. R. L. Bond, *Nature (London) 200*(4913), 1313, December 28, 1963.
32. Y. Kawana, *Chem. Econ. Eng. Rev. 4*(1), *45,* 13 (1972).
33. R. E. Gannon and V. Krukonis, Arc-coal process development, *R&D Report* No. 34, *Final Report,* Office of Coal Research (Contract No. 14-32-0001-1215, November 30, 1971).
34. Office of Coal Research, Review and evaluation of 300 MM lbs./yr. acetylene plant-AVCO arc-coal process, *R&D Report* No. 67: *Final Report,* (Contract No. 14-32-0001-1215, November 30, 1971).
35. *Chem. Market Rep.,* Acetylene from coal could replace olefin as a major chemical feedstock, AVCO says, January 19, 1981, p. 7.
36. *Chem. Eng.* New techniques brighten outlook for plasma, June 1, 1970, p. 56.
37. J. Happel and L. Kramer, *Ann. N.Y. Acad. Sci. 147,* 592 (1969); U.S. Patents 3,156,733 (1964); 3,156,734 (1964); 3,227,771 (1964).
38. D. F. Othmer, *Hydrocarb. Proc. 44*(3), 145 (1965).
39. *Chem. Tech.,* August 1976, p. 477.
40. KTH wet-wall reactor, *Kemisk Tidskrift (Sweden),* No. 1-2 (1973); Swedish Patent 7214094-0 (February 17, 1975).
41. A. C. McKinnis, *Ind. Eng. Chem. 47* 850 (1955).
42. R. J. Tedeschi, Process design development, *Ind. Eng. Chem., 7,* 303 (1968).
43. M. J. Barry, U.S. Patent 3,069,247 (December 18, 1962).
44. H. H. Sisler, *Chemistry in Non-Aqueous Solvents,* Reinhold, New York, 1951, pp. 19-50.
45. *Petrol. Ref.,* Acetylene-Wulff Process Company, November 1959, p. 207; September 1953, p. 162 and references cited therein.

46. *Petrol. Ref.*, Acetylene (SBA-Kellogg)-the M. W. Kellogg Co., November 1959, p. 203.
47. A. Kent, *Acetylene*, (Painting by A. Kent), presented by the Bjorksten Laboratories for Sponsored Industrial Research; gift to the author by Dr. J. Bjorksten.
48. J. W. Copenhaver and M. H. Bigelow, *Acetylene and Carbon Monoxide Chemistry*, Reinhold, New York, 1949, pp. 246-294, 310-345.
49. I. Pasquon, *Chemica Ind. 43*, 396 (1961).
50. R. J. Tedeschi and G. L. Moore, *Ind. Eng. Chem.*, Product Research Dev., 9, *83 (1970).*
51. *R. J. Tedeschi and G. L. Moore,* J. Org. Chem. 34, *435 (1969).*
52. *M. G. Erskine,* Chemical Economics Handbook, Acetylene, *605.5020,* A-Z; *605.5021* A-H, April 1968; 605.5021 *A-V,* August 1975, Stanford Research Institute (SRI), Menlo Park, Calif.
53. *Chem. Market. Rep.*, Acetylene, chemical profile, July 1, 1968, p. 9.
54. V. G. Hahn, *The Petrochemical Industry*, McGraw-Hill, New York, 1970, p. 19.
55. *Chem. Market. Rep.*, *Acetylene*, chemical profile, April 5, 1971; November 8, 1976, p. 9.
56. L. J. E. Hubou, German Patent, 103,862 (June 27, 1899).
57. C. Kaufman and R. H. Hall, U.S. Patent 2,453,440 (November 9, 1948); 2,493,841 (December 27, 1949).
58. E. P. Schoch, *Oil Gas J. 43*, 63 (1945).
59. A. Frank, DPR Patents, 103,862 (1898); 112,416 (1899).
60. R. E. Kirk and D. F. Othmer, *Encyclopedia of Chemical Technology*, *Vol. 4*, 3rd ed., Interscience, New York, 1978, p. 652.
61. *PB Reports*, PB 1345; PB 11396; PB 27713 (BIOS 355), U.S. Department of Commerce, Washington, D.C., 1945.
62. J. W. Reppe, *Justus Liebigs Ann. Chem. 1*, 582 (1953).
63. R. J. Tedeschi, *Ann. N.Y. Acad. Sci. 214*, 50-52 (1973).
64. M. Salkind and E. H. Riddle, *Ind. Eng. Chem. 51*, 1232, (1959); *51*, 1328, (1959).
65. H. T. Neher, E. H. Specht and A. Neuman, U.S. Patent 2,582,911 (January 15, 1952); 2,773,063 (December 4, 1956).
66. C. R. Adams and T. J. Jennings, *J. Catal. 2*, 63 (1963).
67. *Hydrocarb. Proc.*, Acrylic esters-Toyo Soda Manufacturing Co. Ltd., *48*(11), 145 (November 1969), 152-154 (May 1969).
68. D. W. McDonald, K. M. Taylor, and D. M. Brown, *Petrol. Ref. 40*(7), 145-154 (1961).

69. J. D. Idol, Jr., U.S. Patent 2,904,580 (September 15, 1959).
70. *Chem. Week 88*(4), (1961), p. 39.
71. *Oil, Paint and Drug Rep.*, Monsanto completing its phase-out of acetylene-based petrochemicals, September 28, 1970; *Chem. Eng. News*, September 28, 1970, p. 17.
72. F. Veatch, J. L. Callahan, J. D. Idol, Jr., and E. C. Milberger, *Chem. Eng. Prog. 56*(10), 65 (1960).
73. F. Veatch, J. L. Callahan, and J. D. Idol, Jr., *Hydrocarbon Proc. 41*(11), 187 (November 1962).
74. J. L. Blackford, *Chemical Economics Handbook, Acrylonitrile*, 607.5030 *A-Z*, December 1969, Stanford Research Institute (SRI), Menlo Park, Calif.
75. Petrochemical handbook, *Hydrocarbon Proc. 50*(11), 121 (November 1971).
76. G. Corporali, *AIChE Sympos.*, Ser. 69, 127 (1973).
77. R. B. Stoubaugh, S. G. McH. Clark and G. D. Camirand, *Hydrocarbon Proc. 50*(1), 109-120 (January 1971).
78. I. K. Kolchin, *Khim. Prom. 49*(11), 815 (1973).
79. P. R. Pujado, B. V. Vora, and A. P. Krueding, *Hydrocarbon Proc. 56*(5), 169-172 (May 1977).
80. W. H. Carothers, A. M. Collins, and J. E. Kirby, *J. Am. Chem. Soc. 53*, 4203 (1931); U.S. Patent 1,950,432 (March 13, 1934); 2,227,478 (January 7, 1941).
81. R. E. Kirk and D. F. Othmer, *Encyclopedia of Chemical Technology, Vol. 5*, 3rd ed., Interscience, New York, 1978, pp. 156-170, 188-204, 224-227 and references cited therein.
82. Vinyl Acetylene, Br. Patent 401,678 (November 13, 1933).
83. F. B. Downing, U.S. Patent 1,950,434 (March 13, 1934).
84. M. Broadbooks, U.S. Patent 2,923,748 (February 2, 1960).
85. R. E. Kirk and D. F. Othmer, *Encyclopedia of Chemical Technology, Vol. 5*, 2nd ed., Interscience, New York, 1964, pp. 215-228.
86. *Encyclopedia of Polymer Science and Technology, Vol. 3*, Interscience, New York, 1965, pp. 705-709.
87. Chloroprene patents: U.S. 2,391,827 (December 25, 1945); 2,446,475 (August 3, 1948); 2,524,383 (November 3, 1950); 2,879,311 (March 24, 1959).
88. M. G. Erskine, *Chemical Economics Handbook, Butadiene*, 620.5020 *S*, January 1969, Stanford Research Institute, Menlo Park, Calif.
89. M. G. Erskine, *Chemical Economics Handbook, Synthetic Elastomers*, 525.323013, February 1970, Stanford Research Institute (SRI), Menlo Park, Calif.

90. *Chem. Market. Rep.*, Neoprene, May 22, 1978.
91. C. Weizmann, *J. Chem. Soc. London*, 2841 (1953).
92. H. S. Taylor and W. J. Shenk, *J. Am. Chem. Soc. 63*, 2756 (1941).
93. R. J. Tedeschi, U.S. Patent 3,082,260 (March 19, 1963).
94. *Chem. Eng. News*, Isoprene technology faces new influences, May 17, 1965, p. 50.
95. *Chem. Eng. News*, Isoprene makers face go, no-go decision, April 3, 1967, p. 60.
96. M. DeMalde, A. Balducci and V. Cariati, *Hydrocarb. Proc. Petrol. Ref. 43*, 149 (1964).
97. M. DeMalde, *Chim. Ind.*, 665-672 (June 1963).
98. *J. Commerce*, ANIC building rubber facility, June 4, 1970.
99. *Chem. Eng.*, New bloom for isoprene, December 14, 1970, p. 90.
100. A. Heath, High purity isoprene from acetone acetylene, *Chem. Eng.*, October 1, 1973, pp. 48-49.
101. R. E. Kirk and D. F. Othmer, *Encyclopedia of Chemical Technology, Vol. 12,* 2nd ed., Interscience, New York, 1967, pp. 73-79.
102. Br. Patents: 841,746 (July 20, 1960); 841,748 (July 20, 1960).
103. *Hydrocarb. Proc. Petrol. Ref.*, IFP isoprene process, *42*, 187 (1963).
104. A. A. DiGiacomo, J. B. Maerker, and J. W. Schall, *Chem. Eng. Progr.* 57(5), 35 (1961).
105. S. A. Miller, *Chem. Ind.*, 4-16 (January 5, 1963).
106. German Patents: 174,068 (July 28, 1904); 733,750 (March 4, 1943).
107. German Patents: 171,900 (July 26, 1905); 208,834 (September 28, 1907).
108. German Patents: 263,457 (August 14, 1912; 464,320 (December 15, 1925).
109. Swiss Patent 293,110 (January 10, 1944); German Patent 464,320 (December 5, 1925).
110. M. G. Erskine, *Chemical Economics Handbook, Trichloroethylene*, 697.3030 A-K, August 1968.
111. S. A. Miller, Tri and perchloroethylene, *Chem. Proc. Eng.*, June 1966, pp. 269-275.
112. M. G. Erskine, *Chemical Economics Handbook, Perchloroethylene* 685.5030 *A-K*, August 1968.
113. *Chem. Eng. News*, Wide scope seen for dry cleaning chemicals, December 11, 1967, pp. 30-31.

114. L. F. Albright, *Chem. Eng.*, March 27, 1967, pp. 123-130; April 10, 1967, pp. 219-226.
115. L. I. Nass, *Encyclopedia of PVC, Vol. 1*, Marcel Dekker, New York, 1976, pp. 15-32 and references cited therein.
116. Vinyl Chloride Catalysts: U.S. Patent 1,919,886 (1933); Br. 339,727 (1928); 339,093 (1928).
117. R. E. Lynn and K. A. Kobe, *Ind. Eng. Chem. 46*, 633 (1954).
118. R. R. Reese, U.S. Patent 2,601,322 (1952).
119. K. Washimi and T. Asakura, *Chem. Eng.*, October 24, 1966, pp. 133-138; November 21, 1966, pp. 121-126.
120. S. Gomi, *Hydrocarb. Proc. 43*(11), 165-167 (November 1964).
121. J. A. Buckley, *Chem. Eng.*, November 21, 1966, pp. 102-104.
122. F. F. Braconier and J. A. Godart, U.S. Patent 2,779,804 (January 29, 1957).
123. Vinyl Chloride: Br. Patent 954,791 (April 8, 1964).
124. L. F. Albright, *Processes for Major Addition-type Plastics and their Monomers*, McGraw-Hill, New York, 1974, pp. 173, 211, 228-230.
125. R. B. Stoubaugh, W. C. Allen, Jr., and Van R. H. Sternberg, *Hydrocarb. Proc. 51*, (5) May 1972, pp. 153-161.
126. R. E. Kirk and D. F. Othmer, *Encyclopedia of Chemical Technology, Vol. 21*, 2nd ed., Interscience, New York, 1964, pp. 321-327.
127. *Petrol. Ref.*, Vinyl acetate, *38*(11), 304 (November 1959).
128. R. P. Arganbright and R. J. Evans, *Hydrocarb. Proc. 43* (11) November 1964, pp. 159-163 and references cited therein.
129. *Hydrocarb. Proc. 46*(4), April 1967, pp. 146-149.
130. R. E. Robinson, U.S. patent 3,190,912 (June 22, 1965).
131. Koon Ling Ring, *Chemical Economics Handbook, Product Review on Acetylene, Olefins*, 300.5000 *A-V*, and references cited therein, October 1978, Stanford Research Institute (SAI), Menlo Park, Calif.
132. *Chem. Market Rep.*, Acetylene, chemical profile, October 1, 1979, p. 9.
133. *Chem. Eng.*, Coal technology reigns at AIChE gathering, January 1, 1979, pp. 46-49.
134. *Chem. Eng. News*, Coal gasification unit okayed in Illinois, January 21, 1980, p. 8.

135. *The New York Times*, The synthetic solution: The rub is in the cost, July 15, 1979, Sec. 3, p. 1.
136. *Chem. Eng. News*, Synfuels: Unknown and costly option, August 27, 1979, pp. 20-28.
137. *Chem. Eng. News*, 1980s Energy outlook: Gloom and doom, January 21, 1980, pp. 40-41.
138. *Chem. Eng. News*, Underground coal gasification: A crucial test, December 3, 1979, pp. 19-29.
139. *Chem. Eng. News*, Fluidized-bed combustion unit operational, November 26, 1979, pp. 24-25.
140. *Chem. Eng. News*, South Africa commits to oil-from-coal process, September 17, 1979, pp. 13-16.
141. *Chem. Eng. News*, Eastman to make chemicals from coal, January 14, 1980, p. 6.
142. *Chem. Eng. News*, Study sees need for diverse energy sources, January 28, 1980, pp. 39-40.
143. P. Lewis, France sets 2 breeder reactors, *The New York Times*, February 27, 1980, pp. D-1, D-4.
144. P. Lewis, 66 nations urge fast-breeder reactor. *The New York Times*, March 1, 1980, pp. 27-28.

2

SPECIALTY ACETYLENE CHEMICALS AND DERIVATIVES

2.1 REPPE TECHNOLOGY

The first commercial development of acetylenic chemicals has its origins in the achievements and legend of Julius Walter Reppe [1-4]. American and British military-civilian teams were the first to bring back details of a new acetylene-based technology by personal interviews with Dr. Reppe after the defeat of Germany in World War II. A critical shortage of key war materials such as rubber and oil was the motivating force for some of the innovative and hazardous approaches undertaken by Reppe and coworkers. The details of this pioneering work and its industrial applications have been well documented [4,13].* Reppe's research was started in the early 1930s and reached its zenith and commercial fruition by 1940.

In essence, Reppe technology was based on the use of calcium carbide acetylene and its reactions under pressure with compounds such as alcohols, acids, carbon monoxide, and carbonyl compounds. The ethynylation reaction leading to butyndiol, and eventually to the butanediol-butadiene process for synthetic rubber, was important in keeping the German Wehrmacht rolling as the Allied blockade became more effective in the later years of the war.

$$OHCH_2\text{-}C\equiv C\text{-}CH_2OH \xrightarrow{H_2} OHCH_2CH_2CH_2CH_2OH \xrightarrow{-H_2O}$$

$$CH_2{=}CH\text{-}CH{=}CH_2$$

*The Copenhaver volume cites many American and British military government reports (PB,FIAT,BIOS) as its sources (see Chaps. 2, Vinylation and 3, Ethynylation).

Typical of the speed with which Reppe technology was utilized in the war effort was the scale-up of a 300 t/month pilot plant at Schkopau to a 2500 t/month plant at Ludwigshafen within 2 years. The safe transport and handling of acetylene under pressure to prevent explosions and detonations was a critical problem that was successfully solved for the large scale production of butyndiol.

Today, the term *Reppe chemicals* signifies principally acetylenic products such as propargyl alcohol (1-propyn-3-ol) and butyn-1, 4-diol, besides derivatives such as 2-pyrrolidone, N-vinyl-2-pyrrolidone, polyvinylpyrrolidone, vinyl ethers, and tetrahydrofuran. In the United States, the General Aniline and Film Corporation (GAF) is the largest manufacturer of Reppe chemicals, while Badische Aniline Soda Fabrik (BASF) is dominant in Germany. However, since 1969-1970 the E. I. du Pont de Nemours & Company has been operating an acetylene-based tetrahydrofuran plant based on the butyndiol-butanediol route with a reported capacity of about 80 million lb/yr. Such a facility could make du Pont a formidable United States, and world, competitor in Reppe chemicals, should it so choose.

In Table 2-1 are summarized typical applications for Reppe products. The technology, markets, and applications for these versatile specialty chemicals are discussed in greater detail in Secs. 2-2 to 2-12.

2.2 PROPARGYL ALCOHOL-BUTYNDIOL ROUTES

The process operated by GAF is a continuous, fixed-bed, liquid-phase system which produces mainly butyndiol (BD) together with propargyl alcohol (PA). The system is preferentially run in the aqueous phase using formaldehyde and acetylene as reactants over a copper acetylide catalyst deposited on silica gel.

$$\mathrm{CH{\equiv}CH + CH_2O \xrightarrow{CuC_2\text{-}SiO_2} \underset{PA}{H\text{-}C{\equiv}CCH_2OH} + \underset{BD}{OH\text{-}CH_2\text{-}C{\equiv}C\text{-}CH_2OH}}$$

The process is best operated as a complete liquid phase so that the dangerous accumulation of acetylene vapor is avoided. Reaction temperatures as high as 125°C and hydrostatic pressures in excess of 1000 psig have been claimed (Ref. 4, pp. 93-106 and Ref. 5). Under optimum conditions the exothermic reaction is

Table 2-1 Applications for Reppe Chemicals

Product	Applications
Propargyl alcohol	Metal treatment, corrosion inhibition, oil-well acidizing, electroplating, intermediate for Vitamin A and pesticides
Butyn-1,4-diol	Important Reppe starting material, metal treatment, acid pickling, electroplating additive, stabilization of chlorinated solvents, intermediate for pesticides
Butene-1,4-diol	Intermediate for pesticides, pharmaceuticals, fungicides and bacteriacides, polyurethane intermediate
Butane-1,4-diol	Important Reppe intermediate for THF, butyrolactone and polyvinylpyrrolidone polymers; manufacture of polybutylene terphthalate; polyurethanes, Spandex fibers; specialty plasticizers
γ-Butyrolactone	Intermediate for pyrrolidone and polyvinylpyrrolidone(PVP); acetylene solvent (Sacchse); specialty solvent for polymers, lacquers, paint removers, and petroleum processing; intermediate for herbicides, azo dyes, methionine
Tetrahydrofuran	Continuous top-coating of automotive vinyl upholstery, general solvent for resins; coating of cellophane with vinylidene polymers; polymers; polyurethane polymers; Spandex fibers
Pyrrolidone	Intermediate for N-vinyl and PVP polymers, powerful solvent for resins, acetylene solvent, polar reaction solvent, formulating agent, floor waxes, specialty inks, reaction intermediate
N-Methylpyrrolidone	Commercial acetylene solvent, uses same as pyrrolidone, selective extraction solvent for butadiene purification (cracked naphtha), spinning synthetic fibers, surface coatings, pigment dispersant, formulating agent

Table 2-1 (Continued)

Product	Applications
N-Vinylpyrrolidone	Intermediate for PVP polymers; functional monomer; lube oil manufacture; copolymer applications; paints, paper-coating adhesives, cosmetics; increased dye receptivity, pigment dispersant

believed to be operable at 90 to 100°C and pressures of about 5 atm. In aqueous systems the ratio of butyndiol to propargyl alcohol is about 9:1 at a conversion level (based on formaldehyde) of 95%. Even when operating at higher acetylene-to-formaldehyde ratios, the principal product is still butyndiol. This, however, is a fortunate circumstance, since the diol is of much greater commercial importance than propargyl alcohol and is used mainly in the manufacture of tetrahydrofuran, N-methyl-2-pyrrolidone, and polyvinylpyrrolidone.

In "Reppe" days, the process was often plagued by the undesirable formation of acetylene (cuprene) polymers, which clogged the reaction system and lowered productivity greatly. The combination of cuprene polymers with copper acetylide was considered dangerous, particularly during plant shutdown, and was a potential source of fires or explosions. Mercury, bismuth, and iodine compounds were found to be effective in preventing cuprene formation and stabilizing the copper acetylide catalyst. Since GAF and BASF production units have been in successful operation for over 30 years, the explosion and fire hazards in the process appear to have been safely controlled.

Also, in recent years improved catalyst compositions have been developed that are characterized by the use of lower and safer acetylene pressures, besides the formation of fewer cuprene polymers. GAF [6] has developed an improved cuprous acetylide catalyst deposited on a magnesium silicate carrier and containing bismuth as a cuprene inhibitor. The catalyst was shown to be safe against fire or explosion. The ethynylation of 37% formaldehyde could be accomplished in high yields at temperatures as low as 60°C and pressures below 2 atm.

2.3 PROPARGYL ALCOHOL (2-PROPYN-1-01) (PA)

Moore [7] has described an interesting low-pressure process for the preparation of propargyl alcohol in conversions up to 53%, and butyndiol averaging only 26%. The method utilizes dimethylformamide (DMF) as the activating solvent for acetylene, and paraformaldehyde as reactant at temperatures of 85 to 145°C and pressures of 50 to 350 psig. As the paraformaldehyde is gradually depolymerized on heating to anhydrous formaldehyde, it reacts rapidly with excess solvated acetylene in the presence of copper acetylide on carbon. After use, this catalyst is stable to air exposure and mechanical shock and yields essentially no by-product polyacetylene (cuprene) polymer.

McKinley and coworkers [8] developed a continuous process utilizing an acetylene (24%)-acetone solution with aqueous formaldehyde at 95 to 125°C and 1000 psig with copper acetylide on silica as catalyst to give improved conversions to propargyl alcohol. The high hydrostatic pressure employed prevented the separation of acetylene as a gas phase. By varying the formaldehyde-to-acetylene mole ratio from 1.67 to 0.80, the weight ratio of PA to BD varied from 0.3 to 3.3. The total conversion to acetylenic products varied from 22 to 53%.

The above processes for propargyl alcohol show that the use of polar solvents markedly increase alkynol formation. Similar results have been reported in the use of liquid ammonia [9] and other solvents [10] for the preparation of secondary and tertiary alkynols. At present, no commercial solvent-based process is employed for propargyl alcohol, since current market needs appear to be satisfied by the continuous aqueous route.

2.4 BUTYNDIOL (2-BUTYN-1,4-DIOL) (BD)

The processes [11-17] utilized in Germany and the United States employ a liquid feedstock comprising make-up aqueous formaldehyde, recycle butyndiol, propargyl alcohol, and formaldehyde. Acetylene gas is generally introduced concurrently with the feedstock down through the copper acetylide catalyst bed. Temperature control in the reactor towers is of particular importance, since the reaction is exothermic (55 kcal/mol), and too high a temperature or reaction hot spots leads to the rapid formation of cuprene, in spite of inhibitors (mercury, bismuth, iodine compounds). This in turn leads to rapid plugging of the reactors, the buildup of excess back pressure, low reactor productivity, and early plant shutdown. The optimum temperature range is probably between 95 and 125°C.

The reactors used in Germany [11,12] were nearly 60 ft high and 5 ft in diameter and had an empty weight of about 70 t. They were designed to withstand slow acetylene decompositions up to about 3000 psig. Detonations, often in excess of 10,000 psig, were considered impractical to contain.

The copper acetylide catalyst [11,14,16] is prepared in situ in the reactors by impregnating a siliceous catalyst support with a copper salt. A dilute solution of formaldehyde and acetylene, passed through the bed, effects reduction and conversion to copper acetylide. This technique avoids exposure of the shock and oxygen-sensitive cuprous acetylide to handling in the atmosphere. However, this hazard looms during plant shutdown, when reactors and reaction systems have to be cleaned out.

Typically, the reactor effluent prior to separation contains 30 to 33% butyndiol, 0.5 to 4% propargyl alcohol, 1% methanol, 0.5 to 1% formaldehyde, and small amounts of high boilers. Yields of over 90% based on formaldehyde and 80% on acetylene are obtained under optimum process conditions [11-13]. The BD-PA mixture is readily worked up by fractional distillation in which propargyl alcohol, due to favorable vapor-liquid equilibrium, is concentrated to about 35 to 40% as overhead distillate. The BD still-bottoms are further concentrated, and by either crystallization or vacuum distillation pure diol can be obtained. The aqueous solution can also be used and is available commercially.

The aqueous propargyl alcohol overhead distillate is converted to 99.9% product by multistage azeotropic distillations. Treatment with acidified methanol removes certain reaction impurities as overhead, leaving behind the higher boiling propargyl alcohol azeotrope, which is then treated with benzene to yield an enriched (>90%) propargyl alcohol. This material in a final distillation can be upgraded to pure, anhydrous product.

A simplified flow diagram of the butyndiol-propargyl alcohol process is shown in Fig. 2-1. An overall process flow diagram [76] for all Reppe products is shown in Fig. 2-2.

There are no alternative non acetylene processes at present, based on such feedstocks as ethylene, to prepare butyndiol or propargyl alcohol. Also, the possibility of such processes does not seem likely, since an olefin-based route to prepare acetylenics would involve inefficient steps such as dehydrochlorination through a dichloro intermediate or dehydrogenation of the olefin. The BD derivatives, THF and butanediol, however, can be manufactured by alternate routes (see Secs. 2.7.2, 2.8, and 2.8.1.)

2.4.1 Butyndiol (BD) Producers and Markets

The three largest producers of butyndiol currently are BASF, GAF, and du Pont. BASF, inheritor of Reppe-I. G. Farben tech-

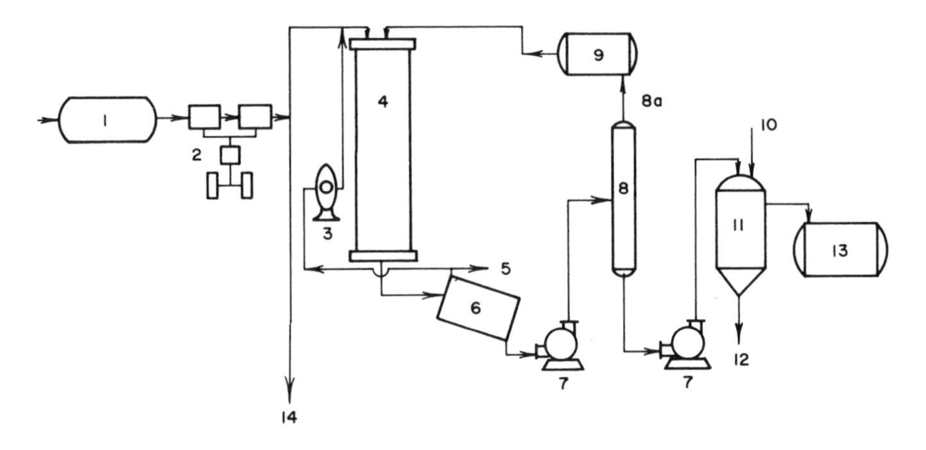

1. Acetylene storage
2. Compressor
3. Recycle compressor
4. Pressure reactor
5. Vent
6. Separator
7. Pumps
8. Stripping still

8a. Recycle formaldehyde
9. Formaldehyde storage
10. NaOH feed
11. Settling tank
12. Waste effluent
13. Butynediol storage
14. Recycle C_2H_2 stream for vinyl pyrrolidone production

Figure 2-1 Flow diagram of butyndiol-propargyl alcohol process. (Excerpted by special permission from *Chemical Engineering*. June 1951. Copyright 1951, McGraw-Hill, Inc. New York.)

nology, resumed manufacture of Reppe products at the end of World War II at Ludwigshafen, Germany. GAF, while still under the jurisdiction of the Alien Custodian Act in the United States, began the manufacture of butyndiol and related Reppe chemicals at Calvert City, Kentucky in 1956. Acetylene, derived from calcium carbide and supplied by Air Reduction Company "across the fence" to GAF, was and still is a part of a well-integrated production facility.

About 10 years later an additional production facility was started by GAF at Texas City, Texas to meet the growing demands for its Reppe line and to utilize cheaper petrochemical acetylene. This plant was operational in the period 1968-1969. Overseas shipments of Reppe chemicals from both the Calvert City and Texas City locations are now made from ports at Houston and Galveston, in preference to New York harbor, which had been

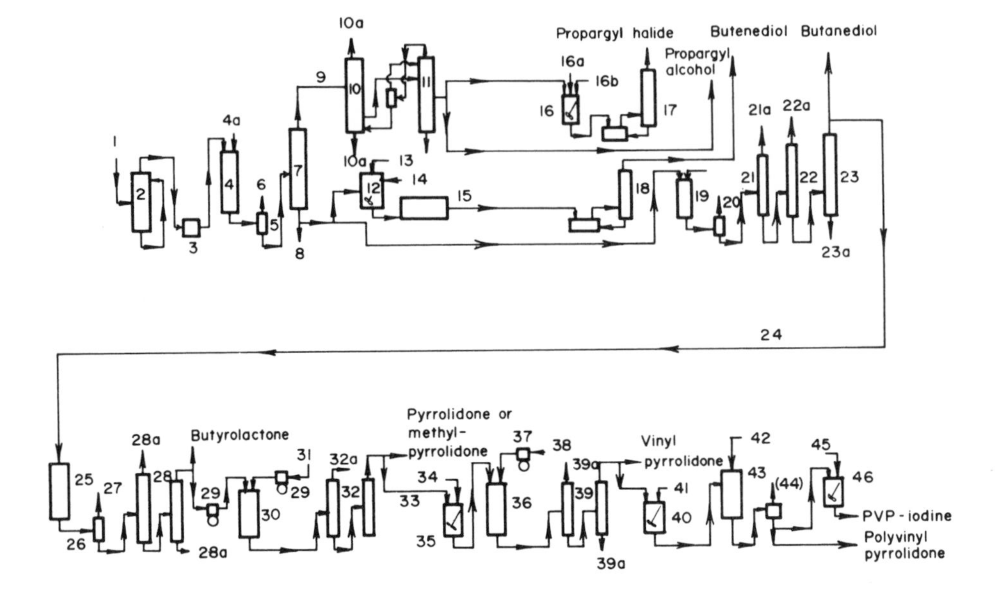

1. Crude acetylene
2. Scrubber
3. Compressor
4. Ethynylation reactor
4a. Formaldehyde
5. Separator
6. Vent
7. Still
8. Butynediol
9. Crude propargyl alcohol
10. Propargyl alcohol still
10a. To waste
11. Propargyl alcohol still
12. Batch hydrogenation
13. Hydrogen feed
14. Catalyst feed
15. Catalyst filter
16. Halogenation reactor
16a. Catalyst
16b. Halogen
17. Propargyl halide still
18. Butenediol batch still
19. Butanediol hydrogenator
20. Vent
21. Butanediol still
21a. To butanol recovery
22. Butanediol still
22a. Waste
23. Final butanediol still
23a. Waste
24. To dehydrogenation reactor
25. Dehydrogenation reactor
26. Separator
27. To hydrogen recovery
28. Butyrolactone stills
28a. Waste
29. Pump
30. Pyrrolidone product reactor
31. Liquid NH_3 or methylamine feed
32. Product stills
32a. Waste
33. Pyrrolidone feed
34. Catalyst feed
35. Catalyst preparation
36. Vinylation reactor

used since mid-1950. GAF is also producing butyndiol and derivatives at Marl, West Germany in a joint venture operation with Chemische Werke Huels AG.

BASF expanded its Reppe business into the United States by the acquisition of Wyandotte Co. in the early 1970s. A production facility rated at 55 million lb/yr (butanediol) in 1974 was erected at the Geismar, Louisiana petrochemical complex [18]. The process utilizes petrochemical acetylene. The most recent producer of butanediol in the United States is du Pont, which completed a 80-million-lb/yr tetrahydrofuran (THF) plant at Houston based on petrochemical acetylene supplied by TENNECO. Du Pont is also considering a non-acetylenic route based on acrolein and methyl propanediol (see Secs. 2.7.1 and 2.7.2).

All three manufacturers of BD have announced major expansions that are expected to be completed in 1977-1978. This increased production is expected to add more than 50% to existing capacity. Total butyndiol capacity in 1976 was about 400 million lb, for both the United States and Europe. The growth rate for Reppe products is about 30%/yr [18]. The major uses for butyndiol are its conversion to butanediol and tetrahydrofuran (see Secs. 2.7 and 2.8.)

2.4.2 Derivatives of Propargyl Alcohol and Butyndiol

From propargyl alcohol and butyndiol, Reppe was able to proliferate into a wide variety of different compounds, taking advantage of the reactivity and functionality of these polar molecules. Similar approaches were also used for secondary and tertiary alkynols and alkyndiols. This chemistry has been well documented both in textbooks [3, 4] and review articles [19].

Butyndiol (1) is the main base upon which the commercial growth of Reppe chemicals is derived, as shown:

---

37. Compressor
38. Purified acetylene
39. Vinylpyrrolidone stills
39a. Waste
40. Polymerizer
41. Catalyst and distilled water
42. Purified air
43. Spray dryer
44. Vent
45. Iodine feed
46. Reactor

Figure 2-2 Continuous process for production of Reppe chemicals (GAF). (Reprinted by special permission from *Chemical Engineering*. February, 1960. Copyright 1960, by McGraw-Hill, Inc., New York.)

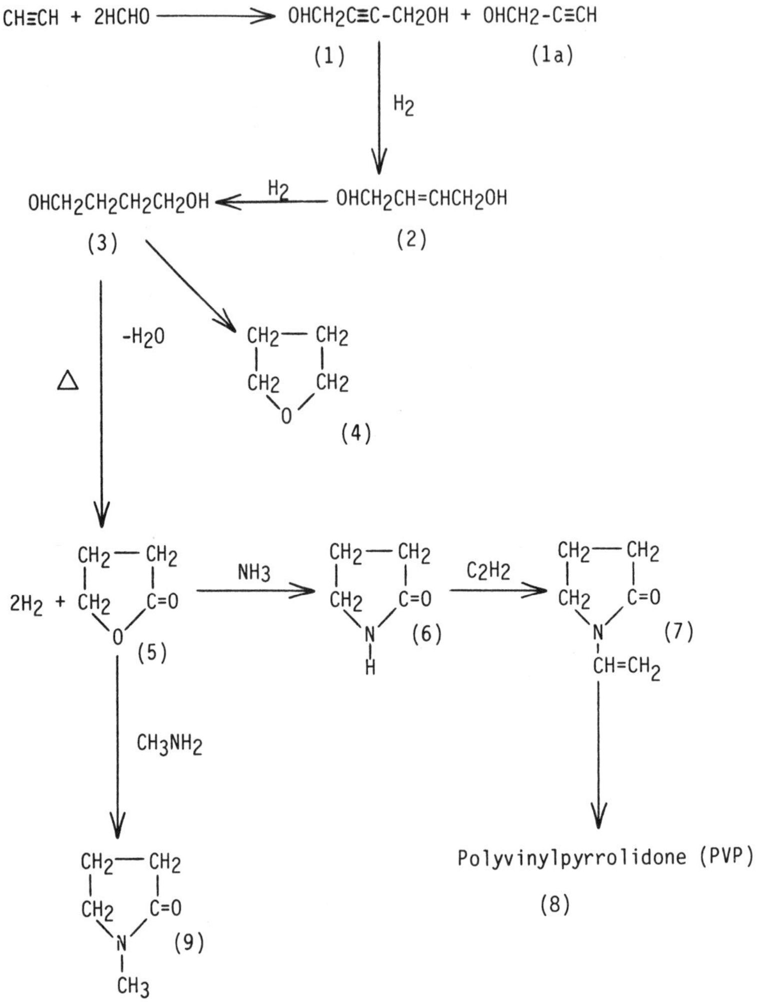

$CH{\equiv}CH + 2HCHO \longrightarrow OHCH_2C{\equiv}C\text{-}CH_2OH + OHCH_2\text{-}C{\equiv}CH$

(1) (1a)

$\downarrow H_2$

$OHCH_2CH_2CH_2CH_2OH \xleftarrow{H_2} OHCH_2CH{=}CHCH_2OH$

(3) (2)

(3) $\rightarrow$ CH_2—CH_2 / CH_2 CH_2 / O (4)

(3) $\xrightarrow{-H_2O,\ \Delta}$ $2H_2$ + CH_2—CH_2 / CH_2 C=O / O (5)

(5) $\xrightarrow{NH_3}$ CH_2—CH_2 / CH_2 C=O / N–H (6) $\xrightarrow{C_2H_2}$ CH_2—CH_2 / CH_2 C=O / N–$CH{=}CH_2$ (7)

(7) $\downarrow$ Polyvinylpyrrolidone (PVP) (8)

(5) $\xrightarrow{CH_3NH_2}$ CH_2—CH_2 / CH_2 C=O / N–CH_3 (9)

Of the above products, tetrahydrofuran (THF) (4), polyvinylpyrrolidone and copolymers (8), N-methyl-2-pyrrolidone (9), butyrolactone (5), and vinylpyrrolidone (7) are probably of greatest commercial importance, although growing specialty markets are enjoyed by the remaining products, particularly butyndiol (1) and butanediol (3). Some of the more important uses for these materials include cosmetics (hair preparations), corrosion inhibitors, electroplate additives, acetylene solvents, polyurethanes, synthetic fibers, lube oil additives, pharmaceuticals, floor polish additives, and beer clarification aids.

2.5 USES AND MARKETS FOR PROPARGYL ALCOHOL (PA) AND BUTYNDIOL (BD)

Both propargyl alcohol and butyndiol, characterized by a high percentage of polar functionality for their respective molecular weights, are versatile, reactive intermediates that are used in a variety of applications. A number of these uses are still embryonic or of relatively modest volume and include uses in such areas as drugs, pesticides, vitamin intermediates, fragrances, metal treatment, and specialty polymers. Larger volume uses, particularly for BD and its derivatives, are discussed in this chapter.

2.5.1 Propargyl Alcohol

This interesting and reactive alkynol has never developed, marketwise, into a large-volume acetylenic. It is, in essence, the little brother of butyndiol and can be considered a by-product of BD manufacture. Since process information is proprietary, it is assumed that the GAF butyndiol-propargyl alcohol process is operated so that alkynol formation is probably less than 10% of theory on a total conversion level of over 90%. Also, a substantial amount of propargyl alcohol is believed to be continuously recycled in the process to aid process economics and produce additional butyndiol.

The use of basic catalysts (hydroxides, alkali metal acetylides) that function well for higher aldehydes and ketones [10, 20] results in very poor yields of either propargyl alcohol or butyndiol due to the instability of both formaldehyde and the acetylenics in alkaline media. A modified Reppe-type process that utilizes anhydrous in situ-generated formaldehyde and acetylene, dissolved in a polar solvent, might be adaptable to continuous operation.

$$C_2H_2 + HCHO \xrightarrow[\text{polar solvent}]{CuC_2\text{-}C} HC{\equiv}C\text{-}CH_2OH$$

By the use of new, more specific and active catalysts, it should be possible to carry out the reaction at lower temperatures (<25°C) with a larger excess of acetylene, which would further favor alkynol formation.

The total U.S. market for propargyl alcohol is believed to be at least 1 million lb, and could be as high as 3 million lb annually. The BASF production in Germany may be higher than the U.S. market. The principal use is in metal-treatment applications and includes corrosion inhibition (oil well acidizing and acid pickling), antihydrogen embrittlement, metal cleaning, and electroplating. It is also used to manufacture propargyl chloride and bromide, which are employed in the manufacture of vitamin A. This use, however, is probably below 100,000 lb/yr.

Propargyl alcohol [21,22] is used extensively in the southwest regions of the United States for oil well acidizing and this use is believed to be at least 1 million lb annually. It is sold in a minimum purity of 97% (straw-colored liquid), although concentrated aqueous material is also available. The latter material, cheaper in price, has also been used for acidizing operations.

2.5.2 Butyndiol (BD)

The principal use for BD [23,24] is as an intermediate in the manufacture of vinyl- and polyvinylpyrrolidones, the end uses being directed to a wide variety of polymer applications (see Sec. 2.10-2.12).

Due to the structural feature of a triple-bond alpha to hydroxyl groups, it functions well in a variety of metal-treating applications such as electroplating, acid pickling, scale removal, and antihydrogen embrittlement. Metal applications probably utilize between 1.5 and 2.0 million lb annually. Ethoxylated butyndiol is also used in electroplating.

Other specialty uses [23] include stabilization of halogenated hydrocarbons [24,25], as an agricultural agent (defoliant [26], insecticide [27], herbicide [28], paint and varnish remover [29], polymerization accelerator [30], and polyurethane intermediate [31,32]. The latter use, in which BD is used as a chain extender for urethane rubbers [31], has been found to increase the tensile and tear strengths of rubbers markedly. Its use in polyester formation is also indicated for both surface coatings and synthetic fibers.

At present, the total U.S. market for BD itself may be about 2 to 3 million lb. Although its greatest importance will continue as a source for other large-volume Reppe chemicals, it is likely that other new and unique uses for this reactive material will be developed based on the retention of the triple bond.

2.6 2-BUTENE-1,4-DIOL

The semihydrogenation of butyndiol to butenediol is a deceivingly simple reaction, which, unlike secondary and tertiary acetylenic diols (Ref. 9, pp. 91-97), proceeds nonselectively to yield a variety of products. Reppe [11,12], (Ref. 4, pp. 131-133) found that nickel, copper, or cobalt catalysts were unsatisfactory due to poisoning and the formation of resins.

$$OHCH_2C{\equiv}CCH_2OH + H_2 \longrightarrow OHCH_2CH{=}CHCH_2OH$$

Iron, derived from iron carbonyl or electrolysis, when used in a continuous system under pressure and under carefully prescribed conditions, was found to give the best results. At 150 to 200 atm and 50 to 100°C, it was possible to obtain butenediol in 93% purity containing about 2% unconverted acetylenic and 5% saturated diol. As the temperature was raised from the initial 50°C range to over 90°C to compensate for diminished reaction rate, the yield to butene diol fell markedly.

Hydrogenation of BD in ethanol with Raney nickel [33] yields the following unselective mixture: 52% cis-butenediol; 14% trans-butenediol; 11% butanediol; 10% butynediol; 9% 2,5-dihydrofuran; and 2% 2-buten-1-ol. Besides the usual olefinic, saturated and acetylenic diols, Reppe (Ref. 1, p. 40) noted in the partial reduction of BD the formation of 4-hydroxybutyraldehyde, 1-butanol, tetrahydrofuran, and furan (see Sec. 2.7).

The process and catalyst currently being employed by GAF has not been disclosed. However, with the significant advances reported [34,35] in recent years in hydrogenation art, and the general acceptance of platinum group (Pd, Rh, Pt) metals or compounds as selective catalysts for the triple bond, the development of new modified catalysts for BD hydrogenation seems likely.

Although reputed to be a tank-car chemical, the total U.S. usage of butenediol is believed to be below 2 million lb (1974) annually. Its inadequate high-temperature stability (dehydrates and polymerized above 165°C) and the fact that it averages only 93% in purity may limit its marketability. The commercial product is mainly the cis isomer and is a clear, amber liquid, very soluble in water and ethanol, but sparingly soluble in benzene.

Specialty uses in the manufacture of insecticides, pharmaceuticals, and resins have been disclosed [24]. A cyclic sulfite insecticide (Thiodan) [36] made by the Diels-Adler addition of butenediol with hexachlorocyclopentadiene followed by reaction with thionyl chloride has been developed by FMC Corp. Halogenated esters of butenediol have been used as fungicides, germicides, and slime-preventing agents in paper manufacture [37]. Butenediol has also been proposed as an electroplate additive [38] to give bright, ductile nickel plate. When reacted with toluene diisocyanate, butenediol is claimed to yield polyurethane that give structural laminates with high flex strength and modulus [39].

2.7 BUTANE-1,4-DIOL (BDS)

The hydrogenation of BD to butanediol (BDS) does not require as careful a control of process conditions as is required for the olefinic precursor. However, it is important that an active, nonacidic catalyst be used so that the vulnerable butenediol stage is passed rapidly. Butyndiol, in the presence of traces of acid, is hydrogenated to γ-hydroxybutyraldehyde, which is converted only slowly and at higher temperatures and pressures to the desired saturated diol. By-products possible in the hydrogenation of BD to butanediol are shown below.

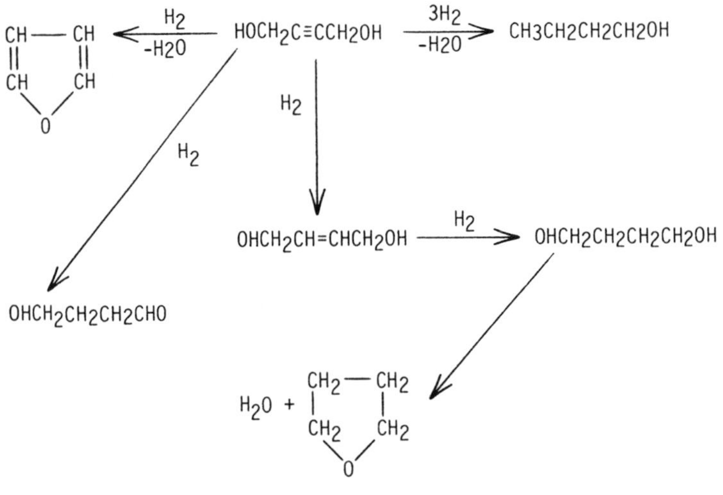

$$\text{CH=CH-CH=CH-O (furan)} \xleftarrow[-H_2O]{H_2} HOCH_2C{\equiv}CCH_2OH \xrightarrow[-H_2O]{3H_2} CH_3CH_2CH_2CH_2OH$$

$$HOCH_2C{\equiv}CCH_2OH \xrightarrow{H_2} OHCH_2CH_2CH_2CHO$$

$$HOCH_2C{\equiv}CCH_2OH \xrightarrow{H_2} OHCH_2CH{=}CHCH_2OH \xrightarrow{H_2} OHCH_2CH_2CH_2CH_2OH$$

$$OHCH_2CH_2CH_2CH_2OH \longrightarrow H_2O + \text{CH}_2\text{-CH}_2\text{-CH}_2\text{-CH}_2\text{-O (tetrahydrofuran)}$$

The original German process (Ref. 3, p. 92, Ref. 4, p. 101) used 35% aqueous butyndiol as feedstock for a continuous hydrogenation tower operated at 70 to 140°C and 300 atm. The catalyst was a nickel (15%)-copper (5%)-manganese (0.7%) composition deposited on a siliceous carrier. With fresh catalyst, the hydrogenation could be carried out at 70°C, but with loss of activity, temperature was graudally raised to 140°C to increase productivity. By operating at constant pressure (300 atm), maximum catalyst life was claimed. Under optimum process conditions, yields of butanediol of 95% of theory were obtained, showing that hydrogenolysis or dehydration by-products (outlined above) were minor.

The present U.S. plants (GAF and Du Pont) [40] for both butyn- and butanediols are believed to use technology similar to that developed by Reppe, particularly for ethynylation. However, it is expected that the hydrogenation step has been modified somewhat, particularly in terms of the catalyst. The development of improved nickel and noble metal (palladium or platinum) catalysts has progressed significantly during the last 30 years, and it is likely that the original Reppe catalyst has been supplanted.

An important use for butanediol is as an intermediate for the manufacture of butyrolactone, 2-pyrrolidone, N-vinyl-2-pyrrolidone, and tetrahydrofuran. These uses in 1966 have been estimated by the author to have resulted in the production of over 30 million lb of the saturated diol. During the period 1970-1974, butanediol production was likely in excess of 50 million lb, since Du Pont [40] was operating a large tetrahydrofuran (THF) plant at Houston (capacity 80 million lb). THF and butanediol are probably used, respectively, in polyurethane and polyglycol ether formation. The latter use for THF is directed to the synthetic fiber Lycra and related Spandex fibers.

Butanediol is employed by polyurethane producers to impart flexibility to the resins, and as such is probably employed with polyglycol ethers obtained from THF. It has similar potential in the manufacture of polyesters from maleic, phthalic, and pyromellitic anhydrides. These uses may be approximately 1 million lb annually.

Butanediol is used in the production of specialty plasticizers (sebacates) for low-temperature applications (refrigerator gaskets). It is also a potentially attractive material for the production of 6,6-nylon. Its 1966 price of 28¢/lb excluded it from this use since it was not competitive with the nylon intermediate 1,4-dichlorobutane derived from butadiene. However, once Du Pont's large plant for butane diol and THF is at full capacity, the diol may be as attractive a nylon intermediate as dichlorobutane. Trade literature (GAF) [24] available for butanediol also describes uses as a chemical intermediate, solvent, cross-linker, plasticizer, and chain extender.

2.7.1 Butanediol (BDS) Outlook and Nonacetylenic Technology

Reppe technology through 1979 is projected to use increased amounts of acetylene for butanediol production. The 1971 usage of acetylene has been estimated at 21 million lb [41] and is expected to grow to 38 million lb by 1979. Acetylene usage for the sister product tetrahydrofuran (THF) in this same period will grow from 34 to 47 million lb.

Total BDS production in 1974 was about 65 million lb (captive use 35 MM lb, merchant sales 30 MM lb). The outlook for 1979 is growth to 120 million lb. About one-half of the 1974 to 1979 increase (55 MM lb) will be due to the production of polybutylene terephthalate (PBT), which is expected to grow from 8 million lb (1974) to 33 million lb (1979). This polymer represents a new growth area for butanediol and is directed mainly to the fiber and tire markets. An increased demand for BDS is also evident in its use as a chain extender for polyurethane and as a source of THF for use in Spandex fibers.

2.7.2 Nonacetylenic Routes for Butanediol

Although Reppe technology has enjoyed impressive growth during the last thirty years, the threat of non-acetylene-based processes is always present. Major reasons for considering alternative routes have been the increasing cost of petrochemical acetylene, its limited supply, and the potential hazards of handling acetylene under pressure (see Secs. 1.5 and 1.7).

The important nonacetylenic routes for the production of butane-1,4-diol from cheap raw materials are:

1. Propylene-acrolein (Du Pont)
2. Butadiene-acetic acid (Mitsubishi, BASF)
3. Benzene-maleic anhydride (Mitsubishi)

Maleic anhydride, commercially available via the catalytic oxidation of cheap benzene or butane, was promoted by Mitsubishi [42] primarily as a process for the production of tetrahydrofuran (THF) with γ-butyrolactone formed as a useful Reppe by-product (see Secs. 2.7.4 and 2.8). However, by varying operating temperatures and pressures, it is possible to produce mainly γ-butyrolactone (10:1 over THF). Hydrogenation [42,43] of the latter with a Ni-Co-ThO_2 on kieselguhr catalyst at 250°C and 100 atm produces butanediol in 98% selectivity and 100% conversion in 6 hr. A recent economic evaluation [43] of this process versus the acetylene, propylene, and butadiene routes has shown it to be not competitive for butanediol production (see Sec. 2.7.4). In the sections that follow, the propylene (2.7.3) and butadiene (2.7.4) routes are discussed.

2.7.3 Propylene-Acrolein Route

Du Pont at present purchases acetylene from TENNECO at Houston for its product requirements. However, the growth of its fiber (polyester) business will require additional capacity, and further expansion by the acetylene route may be less attractive than formerly thought. It is known that Du Pont is considering the manufacture of butane-1,4-diol by a nonacetylenic route [44] (shown below) based primarily on acrolein and methyl propanediol. Since Rohm and Haas (also a user of TENNECO acetylene at this site) is

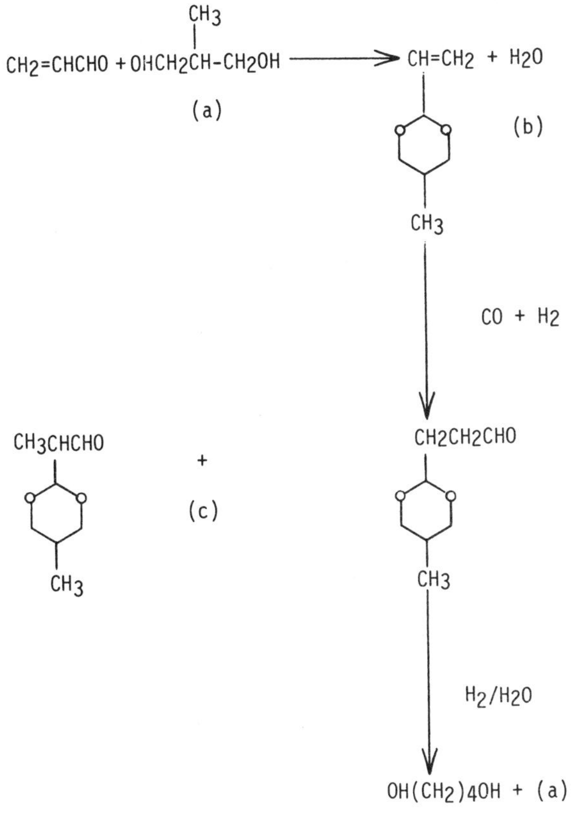

$$CH_2{=}CHCHO + \underset{(a)}{OHCH_2\overset{\overset{CH_3}{|}}{C}H{-}CH_2OH} \longrightarrow \underset{(b)}{CH{=}CH_2} + H_2O$$

(b): $CH{=}CH_2$ attached to a six-membered ring containing two O atoms, with CH_3 at the opposite position

$\downarrow$ $CO + H_2$

CH_2CH_2CHO (ring, CH_3) + (c) CH_3CHCHO (ring, CH_3)

$\downarrow$ H_2/H_2O

$OH(CH_2)_4OH$ + (a)

planning to manufacture acrylates by propylene oxidation to acrolein and acrylic acid, a potentially cheap source of acrolein may be available.

The new Du Pont technology involves the condensation of acrolein with 2-methyl-1,3-propanediol (a) to yield 2-vinyl-5-methyl-1,3-dioxane (b), which on hydroformylation yields an aldehyde mixture of the normal and iso products (c). The normal aldehyde isomer on hydrogenation and hydrolysis yields butanediol and regenerates the methyl propanediol starting material.

The first step of the process (b) proceeds in 87% conversion by simply heating both reactants in a suitable solvent in the presence of a weakly acidic catalyst (polyphosphoric acid) and removing the water formed in the reaction by azeotropic distillation. The reaction of (b) with carbon monoxide and hydrogen (hydroformylation) is carried out under pressure (105 psig) at 110°C using a rhodium carbonyl complex as catalyst. The reaction is rapid, giving a conversion to aldehydes (c) of 97% in 55 min, with a normal-to-iso product ratio of 87:13.

The aldehyde mixture (c) is mixed with 10% acetic acid, hydrolized and then hydrogenated at 100°C and 1000 psig to yield butanediol (98% yield) and methyl propanediol (a) (96% yield). The latter product is recycled to the first step of the process, while butanediol is converted to tetrahydrofuran (THF) by cyclization with sulfuric acid (see Secs. 2.8 to 2.8.4).

Although this process has more steps and utilizes an expensive rhodium catalyst, it appears to be competitive with the Reppe route, primarily due to the very high yields and relatively simple process steps, amenable to continuous operation. It is believed that Du Pont is considering further expansion of its butanediol-tetrahydrofuran facilities via this new technology.

2.7.4 Acetoxylation of Butadiene

A process that in the years ahead could compete effectively with the Reppe acetylene-based process for 1,4-butanediol is a butadiene-based process developed by Mitsubishi Chemical Industries [43,45,46]. The process involves the liquid-phase acetoxylation of butadiene to 1,4-diacetoxy-2-butene, followed by hydrogenation and hydrolysis to butanediol:

$$CH_2{=}CH\text{-}CH{=}CH_2 + CH_3CO_2H + \tfrac{1}{2}O_2 \longrightarrow$$

$$\underset{\text{(a)}}{CH_3CO_2CH_2CH{=}CHCH_2O_2CCH_3} + H_2O$$

$$(a) + H_2 \longrightarrow CH_3CO_2CH_2CH_2CH_2CH_2O_2CCH_3$$

(b)

$$(b) + H_2O \longrightarrow HO\text{-}(CH_2)_4\text{-}OH$$

(c) 1,4-butanediol

The yield in the acetoxylation step is critical to the success of the process. The use of a new palladium-tellurium (Pd-Te) catalyst (deposited on carbon) is claimed to give selectivities to 1,4-diacetoxy-2-butene (a) of about 91%, with isomeric acetoxy-butenes amounting to only 8%. Butadiene conversion is claimed (not substantiated) to be as high as 99% and has been estimated at 9.8 mol/hr g of catalyst. Details of a continuous commercial process have not been disclosed at present. However, in a batch system, reaction is initiated by adding butadiene (200 mmol) and oxygen to acetic acid (200 ml) and catalyst (10 g) at 80°C. Oxygen is introduced at a partial pressure of 3 kg/cm^2. The nitrogen pressure at 80°C is 27 kg/cm^2. An earlier vapor-phase process (Kuraray) gave the diacetoxy (a) product in only 68% selectivity per pass, making it economically unattractive.

The hydrogenation of (a) to the saturated 1,4-diacetoxy butane [45,46] proceeds in 98% selectivity (conversion 100%), using a nickel-zinc catalyst combination on diatomaceous earth. The reaction is carried out at 60 atm hydrogen and 80°C and is complete in 90 min.

The hydrolysis step (most likely acid catalyzed for acetic acid recovery) is believed to be almost quantitative, and overall yields based on butadiene are estimated at about 84%. BASF [47] has also been studying the butadiene-acetoxylation route and has claimed selectivities in the first step (a) of 97% using a Pd_4Te catalyst. The catalyst is prepared by reacting tellurium dioxide (TeO_2) and palladous chloride ($PdCl_2$) in HCl solution, followed by reduction with hydrazine. BASF [48] has also shown that isomeric diacetoxy butenes can be isomerized to the 1,4-diacetate by using platinous chloride ($PtCl_2$) as catalyst. About 51% of the 3,4-diacetoxy isomer can be converted to (a) at 100°C in 60 min using $PtCl_2$. This utilization of by-products has the potential of further reducing process costs.

Brownstein and List [43] have detailed comparative economics for producing butane-1,4-diol at 60 million lb/yr for key processes based on starting materials such as acetylene, propylene, maleic anhydride, and butadiene, based on fourth-quarter 1976

Table 2-2 Comparative Economics of 1,4-Butanediol Processes (60 MM lb plants) Fourth Quarter, 1976

	Reppe (¢/lb)	Propylene (¢/lb)	Maleic Anhydride (¢/lb)	Butadiene (¢/lb)
Raw materials	28.1[a]	29.7[b]	48.6[c]	17.2[d]
Utilities	3.6	2.9	3.4	4.3
Operating costs	2.1	1.9	1.9	2.1
Overhead	6.6	5.9	6.7	6.5
Total	40.4	40.4	60.6	30.1
By-product credit		(1.0)	(7.0)	
Net cost	40.4	39.4	53.6	30.1
RO1 (25%)	8.6	7.4	8.5	9.0
Transfer price	49.0	46.8	62.1	39.1
Fixed capital, $ MM	20.6	17.8	20.3	21.5

[a]Acetylene, 28¢/lb: 37% formaldehyde, 4.8¢/lb.
[b]Acrolein, 28.9¢/lb (propylene, 9.75¢/lb.
[c]Maleic anhydride, 29.6¢/lb (n-butane, 4.8¢/lb.
[d]Butadiene, 18.5¢/lb.

information. The study, summarized in Tables 2-2 and 2-3, indicates a significant advantage for the use of butadiene. If the price of acetylene keeps increasing at a faster rate than butadiene, this could cause serious problems for the Reppe route in the years ahead. The fact that both BASF and Du Pont are actively studying the use of other feedstocks such as propylene (Du Pont) and butadiene (BASF) would indicate that the price gap between petrochemical acetylene and feedstocks such as propylene and butadiene is widening.

The significant difference in transfer price (cents per pound) between the acetylene (49.0¢/lb) and butadiene (39.1¢/lb) processes are summarized in Table 2-2. Also shown (Table 2-3) is the significant price increase in acetylene from 1972 to 1976 and the unfavorable price differential between acetylene-butadiene and acetylene-propylene.

How soon the unfavorable economic projections will affect the Reppe route is unclear. Reppe plants in general are amortized,

Table 2-3 U.S. Price History of 1,4-Butanediol Raw Materials

Year	Acetylene (¢/lb)	Propylene (¢/lb)	Butadiene (¢/lb)	Maleic Anhydride (¢/lb)
1972	11.0	2.9	7.8	13.0
1974	20.0	6.9	14.6	29.0
1976	28.0	9.2	18.5	37.0

Table 2-4 Price Differentials

Year	Acetylene-butadiene	Acetylene-propylene
1972	3.2	8.1
1974	5.4	13.1
1976	9.5	18.8
1980E	11.5	24.5

and oil prices related to propylene and butadiene may be rising faster than natural gas (methane), a common U.S. feedstock for acetylene. Also, coal-based (AVCO) or calcium carbide acetylene might become attractive at a later date (see Secs. 1.2, 1.3.2, 1.5, and 1.7).

2.8 TETRAHYDROFURAN (THF)

Aqueous (35%) butanediol, after stripping to free it of by-product butanol and passing through ion exchangers to eliminate cations (Na^+), can be used directly for the preparation of THF [3,4]. The BD solution is then adjusted to pH 2 with phosphoric acid (0.05-0.1%), pumped through preheaters and finally to tower pressure reactors. The system is operated at 260 to 280°C and pressures of 90 to 100 atm. The yields claimed by Reppe in this process were 99% and the productivity of the process was rated at 4 t THF/m^3 of reactor volume.

$$OH\text{-}CH_2CH_2CH_2CH_2OH \longrightarrow \underbrace{CH_2\text{—}CH_2,\ CH_2\quad CH_2}_{\text{ring closed through O}} + H_2O$$

The initially distilled product is an azeotrope of 94% THF and 6% water, which can be upgraded to anhydrous material by pressure distillation [49].

The details of the Du Pont [40] process have not been divulged but are believed to be similar to the original Reppe method. With the advent of Du Pont as a THF producer (1969), a new, enhanced growth era for Reppe chemicals is possible. The total rated U.S. capacity on the basis of butyndiol for producers such as GAF and Du Pont is probably over 200 million lb. The manufacture of THF is a significant percentage of this total, since the 1969 usage of 37% formaldehyde for BD production has been estimated at 70 million lb [50].

2.8.1 Competitive THF Processes

The original U.S. process for the production of THF was based on the use of furfural, which in turn was based on the acid hydrolysis of polysaccharides (pentosans). The Quaker Oats Co. also used oat hulls at one time (Cedar Rapids, Iowa) as raw material but is now using corn cobs in all three of its plants (Memphis, Tennessee; Omaha, Nebraska; and Cedar Rapids). The total capacity for furfural in 1969 was about 190 million lb. The 1967 consumption was estimated to be in excess of 150 million lb.

Furfural is converted to THF via decarbonylation to furan, followed by hydrogenation [51,52]. Quaker Oats Co. produced it by this route:

```
HC —— CH                        CH —— CH
||    ||        Δ               ||    ||          H2
HC    CCHO  ———————>            CH    CH     ——————————>   THF
  \  /         -CO                \  /
   O                               O
```

The use of steam and a zinc chromite catalyst converts furfural to furan, which is then catalytically hydrogenated to THF. Prior to 1968, Du Pont had used about 40 million lb/yr of furfural for THF production at its Niagara Falls facility [52,53]. However, with the completion of its new THF plant at Houston, this market for furfural will be lost. It is unlikely that furfural will be able to compete with the Reppe route in the years ahead, since chemicals based on natural products (rubber, turpentine, lemon grass oil, etc.) have in the past ultimately proven to be more expensive and less reliable as a source of supply than their synthetic counterparts.

Butadiene, once prepared by the Reppe route in Germany via dehydration of butane-1,4-diol or THF, is now being considered as a competitive raw material for THF manufacture. Petro-Tex has announced a THF process based on butadiene [53]. The process involves the following oxidation-hydrogenation route [54]:

$$CH_2{=}CH{-}CH{=}CH_2 + O_2 \xrightarrow{-H_2O} \text{furan (CH=CH–CH=CH–O ring)} \xrightarrow{H_2} \text{THF (}CH_2CH_2CH_2CH_2O\text{ ring)}$$

Furan is formed by reacting butadiene with oxygen in a 0.5-2:1 mol ratio (O_2:diene) using a molybdenum oxide-bismuth oxide catalyst in the presence of steam. The process technology was licensed [53] to Hodsgoya Chemical Co. (Japan), who are reputed to be building a semiworks plant to evaluate the process. The process, however, has been described as impractical for obtaining high yields [55]. However, butadiene at 19¢/lb is an attractive raw material if overall process efficiencies can be raised to that of the Reppe system.

A THF process based on maleic anhydride has been developed by Kanetaka and coworkers [55] at Mitsubishi Petrochemical Company, Ltd., in Japan. The process involves the following hydrogenation route and is capable of producing γ-butyrolactone, THF, or mixtures of both. The process has also been evaluated for the production of 1,4-butanediol (see Secs. 2.7.2 and 2.7.4) but is not considered competitive [43] with processes based on acetylene or butadiene.

Their publications [55-57] describe reaction parameters, kinetics and control of by-products by the use of a new hydrogenation catalyst. The conversion to THF is greater than 90% per pass, and the conversion on maleic anhydride is 100%. Butyrolactone can be produced as a by-product, if desired. The catalyst has not been disclosed but is believed to be a nickel-precious metal composition [57] that is poisoned by sulfur or carbon monoxide.

Utilizing maleic anhydride (from benzene) at 12¢/lb (1970), a plant of 5000 t/yr operated at full capacity is claimed to produce THF at 29.1¢/lb, giving a pay-out time of 2.2 yr at a THF selling price of 38.0¢/lb. The process is considered to involve lower capital costs than the Reppe route, and since it also produces the intermediate butyrolactone, it is a potential competitor to the acetylene-based route. A large-scale manufacturer of maleic anhydride or benzene might find this process an attractive route, since benzene is a cheap starting material (24-25 ¢/gal).

$$\text{Maleic anhydride} + H_2 \longrightarrow \text{Succinic anhydride (SA)}$$

$$SA + 2H_2 \longrightarrow \gamma\text{-Butyrolactone (BL)} + H_2O$$

$$BL + 2H_2 \longrightarrow \text{Tetrahydrofuran (THF)} + H_2O$$

Mitsubishi Petrochemical [56] intends to license the process in the United States at royalty rates favorable to a potential manufacturer so that it can compete with the Reppe route. Since the Du Pont THF plant at Houston is rumored to have a capacity of 80 million lb/yr and is based on relatively cheap petro-acetylene, it seems the Japanese process would have to be fully integrated with benzene and maleic anhydride production to be competitive. Also, since benzene has now been identified as a potent carcinogen, its use in large quantities may be an unacceptable process hazard.

A new non-acetylene-based process for the production of butane-1,4-diol has been patented [44] by Du Pont and is discussed in Secs. 2.7.1 and 2.7.2. It is possible that Du Pont may use this technology to produce butanediol as starting material for

THF. A butadiene-based process is also considered economically attractive (see Sec. 2.7.4).

2.8.2 Tetrahydrofuran (THF) Production and Applications

Du Pont is the dominant producer of THF in the United States. Its Houston plant in 1969 was rated at 50 million lb/yr, but by 1973 capacity had been expanded to 80-90 million lb annually. The acetylene used for this Reppe-type plant is supplied (pipe line) by TENNECO from its nearby Houston plant. In 1974 this plant used 34 million lb of acetylene, which was equivalent to 80 million lb of THF. By 1979 it was estimated that THF production would grow to 112 million lb annually, requiring 47 million lb of acetylene [41]. The growth rate for THF has been estimated at about 7%/yr.

Although both GAF and BASF-Wyandotte are major suppliers of butanediol, the precursor of THF, neither are considered a significant factor in the production of THF in the United States. GAF, the earliest producer of Reppe chemicals, has directed most of its butyndiol production into other large-volume specialties such as butanediol, γ-butyrolactone, N-methylpyrrolidone, vinyl pyrrolidone and its polymers (PVP and copolymers).

Large amounts of THF are known to be used in the following areas [49,52,53,62,63]:

1. Continuous top-coating of automotive vinyl upholstery (THF-PVC solutions)
2. Coating cellophane with vinylidene copolymers

THF, besides its excellent solvent power for resins, has a high diffusion rate through both PVC and Acrylic resins, which in turn result in increased coating speeds with no tackiness due to solvent retention. Also, uses in protective coatings, film casting, printing inks, adhesives, extraction, and reaction solvent uses, and as a chemical intermediate have been described [49, 62-64].

Another large-volume application for THF is its use as an activating solvent for the manufacture of tetraethyllead (TEL) and tetramethyllead (TML) [58]. The Nalco Chemical Company has developed an electrolytic Grignard route that uses 8000 gal/ electrolytic cells, in which the cell walls are the cathode and lead pellets fed into the top of the cell act as a reacting anode. A mixture of THF and diethylene glycol dibutyl ether functions as the ionizing media for the alkyl Grignard, which under the cell potential is converted to ions, and which react at the anode with the lead pellets to form TEL. The cell reactions are:

Anode

$$4C_2H_5^- + Pb \xrightarrow{-4e^-} (C_2H_5)_4Pb$$

TEL

Cathode

$$4MgCl^+ \xrightarrow{+4e^-} 2Mg + MgCl_2$$

By operating the process with an excess of alkyl halide, the precipitation of magnesium metal is prevented, due to the formation of additional THF soluble grignard.

Overall:

$$Pb + 2C_2H_5MgCl + 2C_2H_5Cl \xrightarrow{EMF} (C_2H_5)_4Pb + 2MgCl_2$$

The total U.S. production of lead alkyls [58] in 1967 was estimated at 624 million lb, with about 70 million lb being exported. TEL was about 80% of the market. Although TML demand increased up to 1970, federal and state activity concerned with air pollution and particularly heavy metals (Hg, Pb, As, etc.) is expected to have very unfavorable consequences upon lead alkyl production in the years ahead [59, 60]. The major producers of TEL are Du Pont, Ethyl Corporation, Houston Chemical, and NALCO Chemical.

2.8.3 THF as a Polymer Intermediate

Perhaps the greatest unrealized potential for THF lies in its versatility as a chemical intermediate [62]. It can be readily transformed into reactive 1,4-dihalobutanes with halogen acids, is readily chlorinated to 2,3-dichlorotetrahydrofuran, and can be polymerized to polyethers, to illustrate but a few of its potentialities.

2.8.4 Spandex Fibers

The polymerization of THF to the polyoxytetramethylene polyol derivative has found recent important applications in the manufacture of Spandex fibers (Lycra) and in polyurethane elastomers.

$$\text{THF (ring: } CH_2\text{—}CH_2\text{—}CH_2\text{—}CH_2\text{—}O\text{)} \longrightarrow HO{+}(CH_2)_4\text{-}O\text{-}(CH_2)_4{+}_n OH$$

Elastomeric or Spandex fibers, due to unique structural features (urethane blocks), possess natural stretch in contrast to mechanically induced stretch as typified by nylon. They were developed primarily as replacements for rubber thread. Spandex fibers were first synthesized by Du Pont in 1947 and introduced commercially in 1959. The fibers can be produced from a variety of diols and diisocyanates to produce elastomers of different properties and end uses.

Du Pont is believed to be producing [62] Lycra mainly from THF (see Secs. 2.8.2 and 2.8.3). It is not known for certain which of the commercially available diisocyanates is being used, but a likely candidate appears to be MDI (diphenylmethane-4,4-diisocyanate).

$$\underset{\text{MDI}}{OCN\text{-}C_6H_4\text{-}CH_2\text{-}C_6H_4\text{-}NCO} + HO{+}(CH_2)_4\text{-}O\text{-}(CH_2)_4{+}_n OH + [OCN\text{-}C_6H_4]_2CH_2$$

$$\downarrow$$

$$\sim OCHN\text{-}C_6H_4\text{-}CH_2\text{-}C_6H_4\text{-}NHCOO\left[(CH_2)_4\text{-}O\text{-}(CH_2)_4\right]_n\text{-}O\text{-}CONH\text{-}C_6H_4\text{-}CH_2\text{-}C_6H_4\text{-}NHCO \sim$$

The patent art has also disclosed polyurethanes from polyethylene glycols and toluene diisocyanate (Du Pont) [65] from adipic acid polyesters of ethylene or propylene glycols with MDI (U.S. Rubber) [66], or from caprolactam (Union Carbide and Polythane) and a suitable diisocyanate [64].

The reaction is generally taken to a prepolymer stage (low-molecular-weight adduct with free OH groups) and later converted to the elastomer (more diisocyanate) either in the spinning bath or a polymerization reactor in a highly polar solvent such as DMF. Diamines, particularly hydrazine, are then used to react with free isocyanate groups to form higher molecular weight linear polyurethane elastomers. The resulting polymers are soluble in THF and can be spun directly from the solvent into the desired fiber filaments. The urea linkages are potential sites for cross-linking with the diamines and are believed to impart the stretch-recovery properties typical of the filaments.

Spandex fibers were originally visualized as a potentially large and rapidly growing market area. This potential has not materialized at present. Production has grown from about 1 million lb in 1961 to about 13 million lb in 1970 [62,64]. The hoped for replacement for other synthetics (stretch nylon and polyesters) has not materialized, since the latter are considerably cheaper than a typical Spandex such as Lycra, which in 1966 averaged $6 to $15/lb (depending on denier). The market for Spandex fibers is still concentrated mainly in women's wear (foundations and swim suits), and until it can be spread, via cheaper manufacturing costs, into other textile uses, it will be limited.

2.9 γ-BUTYROLACTONE (BLO) AND ITS APPLICATIONS [67,68]

Reppe considered the dehydrogenation of butane-1,4-diol to γ-butyrolactone as proceeding via an internal Cannizzaro reaction (Ref. 3, pp. 237,704):

$$HO(CH_2)_4OH \xrightarrow{-H_2} \begin{array}{ccc} CH_2 & — & CH_2 \\ | & & | \\ CHO & & CHO \end{array} \longrightarrow \begin{array}{ccc} CH_2 & — & CH_2 \\ | & & | \\ CH_2OH & & CO_2H \end{array} \xrightarrow{-H_2O} \begin{array}{ccc} CH_2 & — & CH_2 \\ | & & | \\ CH_2 & & C{=}O \\ & \diagdown O \diagup & \end{array}$$

BLO

Copenhaver and Bigelow (Ref. 4, p. 152) suggested an alternate dehydrogenation route based on the known activity and use of copper-chromium catalysts for this type of reaction:

$$OH(CH_2)_4OH \xrightarrow{-H_2} OHCH_2CH_2CH_2CHO \rightleftharpoons \begin{array}{ccc} CH_2 & — & CH_2 \\ | & & | \\ CH_2 & & CHOH \\ & \diagdown O \diagup & \end{array} \xrightarrow{-H_2} BLO$$

The original liquid-phase process [67] was replaced by a more efficient vapor-phase route [4]. This process utilized a copper on pumice catalyst and was operated at 230 to 250°C at atmospheric pressure. Preheated (40°C) butanediol and hydrogen (140-260°C) were vaporized together at about 200°C and passed up through the catalyst bed. Condensed product averaged 90 to 93% BLO, 1 to 3% THF, and 4 to 8% H_2O. The yield of butyrolactone was about 90% of theory. It was readily purified first by atmospheric distillation to remove volatiles (THF and H_2O), followed by vacuum distillation (bp 65°C/4 mm; 203°C, 760 mm).

BLO [68] is manufactured primarily as an intermediate for pyrrolidone, N-methylpyrrolidone, and polyvinylpyrrolidone. Recent process details of the U.S. (GAF) or German (BASF) processes have not been disclosed. Although Reppe's original catalyst was quite durable and had a long life (over 2 yr), it is possible that new dehydrogenation catalysts are currently being used. Significant advances in isomerization and dehydrogenation catalyst technology during the past 30 years in the petroleum industry provide ample evidence for this view.

The 1966 U.S. production of BLO has been estimated by the author at approximately 20 million lb, of which about 14 million lb was believed used for the manufacture of pyrrolidone, vinylpyrrolidone, polyvinylpyrrolidone, and copolymers. Approximately 3 to 5 million lb were used either as a selective acetylene solvent in Sacchse plants or to make N-methyl-pyrrolidone for a similar use. An additional 2 million lb was probably used in a wide variety of specialty solvent applications, particularly for polymers. European production for similar applications may be significantly higher than the U.S. figures, since Reppe chemicals have enjoyed a longer lead time in Europe.

An important large-scale use for BLO via polyvinylpyrrolidone [68] (POLYCLAR AT)* is the use of the latter as an FDA-approved beverage clarification aid. Its use in beer and wine clarification is a potentially important large-volume use since clarity, stability, and flavor of the products are markedly improved. This application, particularly for beer, makes any current (1976-1977) assessment of U.S. BLO production quite difficult, since published data for Reppe chemicals is not available. Clarification of fruit juices is another related application, that has similar growth material.

Butyrolactone [68] has been recommended as a solvent for polyacrylonitrile, Epons, Vinylite NYGL, Dynel, and polyvinylbutyral and as a high boiling component of lacquers. Due to its wide

*POLYCLAR AT: GAF registered trademark for polyvinylpyrrolidone (PVP) used in beverage clarification.

liquid range (-44 to 204°C) and stability, it is a useful solvent for petroleum processing. Its use in paint remover formulations, in the preparation of photographic films, and as an intermediate for herbicides, monoazodyes, methionine, and rocket fuels has also been disclosed [24,68].

2.10 2-PYRROLIDONE AND N-METHYL-2-PYRROLIDONE (NMP)

Butyrolactone reacts readily with either ammonia or methylamine to yield, respectively, pyrrolidone and methylpyrrolidone (Ref. 1, pp. 126-127, Ref. 4, pp. 163-164):

```
                                                  CH2 — CH2
                            NH3                   |      |        + H2O
                  ---------------------->         CH2    C=O
                  |                                 \   /
CH2 — CH2         |                                   N
|      |          |                                   |
CH2    C=O -------|                                   H
  \   /           |
    O             |                               CH2 — CH2
                  |                               |      |        + H2O
                  ---------------------->         CH2    C=O
                           CH3NH2                   \   /
                                                      N
                                                      |
                                                      CH3

                                                      NMP
```

The reaction involves two stages, the first being the opening of the lactone ring to form a γ-hydroxybutyramide derivative, which then cyclizes to the desired pyrrolidone:

```
            CH2 — CH2
CH3NH2  +   |      |    ----->  OHCH2CH2CH2CONHCH3
            CH2    C=O                  |  -H2O
              \   /                     v
                O                      NMP
```

The reaction can be carried out batchwise by reacting butyrolactone with anhydrous ammonia or methylamine at 230°C under pressure [4]. The optimum yield of vacuum distilled product is about 90%.

The process can also be carried out continuously by first forming the hydroxybutyramide precursor at -15°C in a low-pressure reaction. The resulting exotherm raises the reaction temperature to 40°C, at which point the molten amide is pumped to a preheater (180°C) and finally to a two-stage pressure-pipe reactor operated at 250°C and 1200 to 1400 psig. Reppe claimed yields of pyrrolidone of 90 to 95%, and a productivity of 8 t/mo (total reactor volume about 17 liters) using the continuous method.

2.10.1 Uses for 2-Pyrrolidone (2-Pyrol)

Pyrrolidone [24] is used mainly as starting material for the preparation of N-vinyl and polyvinylpyrrolidones (see Sec. 2.11). However, during 1966 about 2 to 3 million lb were sold as a component of styrene-based floor waxes. It has been suggested as a selective solvent for acetylene purification in the Sacchse process (see Sec. 2.9), and it may share some of this market with butyrolactone and N-methylpyrrolidone. GAF has also recommended its use for polymer solutions, for acrylonitrile manufacture, for specialty inks, and as a reaction intermediate [24].

Its use as an intermediate for the manufacture of Nylon 4 fibers has been investigated by a number of companies (3M, GAF, ICI America, Arnold Hoffman) over the past 10 years [69-71].

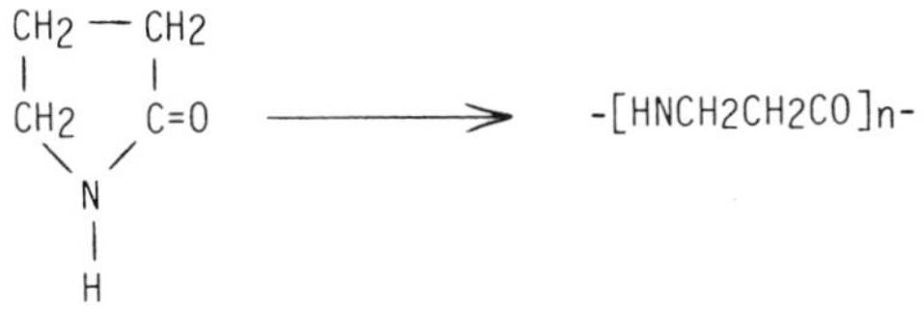

```
CH2 — CH2
|      |
CH2    C=O   ————————>   -[HNCH2CH2CO]n-
  \   /
    N
    |
    H
```

A problem that up to now has prevented its commercialization is the decomposition of the fiber, due to the fact that its decomposition temperature is too close to its melting point. However, in 1968 the Alrac Corporation [69] announced it had overcome decomposition and spinning problems and was manufacturing small quantities of Nylon 4 at Stamford, Connecticut.

The polymer has a melting point of 265°C and, due to its hydrophylic nature, can be readily dyed (direct and vat dyes). It has a moisture regain of 7 to 8% (65% relative humidity, 70°F), which is comparable with cotton and much higher than Nylon 6. Hence, while still having the strength of a polyamide, it also has some of the desirable properties of cellulosic materials, particularly lack of static. The polymer is being considered as a replacement in part for Nylons 6 and 66. Time will determine if these potentials are reached.

2.10.2 Uses for N-Methyl-2-pyrrolidone (M-Pyrol) (NMP) [72]

M-Pyrol,* due to lack of an active hydrogen atom, is not used as a reaction intermediate as is pyrrolidone [24]. Its uses are based on its unique properties (high polarity, complete water miscibility, high solvent capacity, good stability, non-corrosivity, and nonvolatility).

It is used as a selective acetylene solvent (cf. butyrolactone, BLO, Sec. 2.9), and its volume in this application is probably larger than that of BLO. The combined market in 1966 was estimated by the author at about 3 to 5 million lb annually. Polyimide coatings (pyromellitic anhydride-diamine types), used for high-temperature magnet wires, employ M-Pyrol as a solvent. This use was estimated at about 1 million lb annually during the period 1966-1967.

An important large-volume use for N-methylpyrrolidone has developed during the period 1968-1969 in Europe, due to a new BASF process for the production of pure butadiene [73]. Three plants (Erdölchemie, Cologne, West Germany; ICI, Wilton, England; and Industrialimport, Pitesti, Romania) are now producing annually a combined 175,000 t of butadiene from cracked naphtha, using the BASF process.

The process involves a carefully controlled countercurrent extraction system in which a gaseous C_4 mixture fed into the center of a multistage column is scrubbed by a descending stream of NMP effecting a separation of butenes (head of scrubber) from mixed (1,2 and 1,3) butadienes and C_4 acetylenes. The butadienes and C_4 acetylenes are boiled from the solvent, and pure 1,3-butadiene is then obtained by fractional distillation. The process is capable of separating pure butadiene from a minimum of sixteen components, in which the butadiene content in the cracked naphtha stream averages about 45 vol %.

It is quite possible that this technology will be utilized in the United States in the near future, giving a further large growth potential to Reppe products. The method also may have potential for other petrochemical-based products such as isoprene and isobutylene. At present, furfural and curprous ammonium acetate are used as an extraction medium for butadiene in the United States. However, with the natural raw material source (corn cobs)

*GAF registered trademark.

becoming inadequate for furfural production, it would seem this compound will be gradually replaced as a petrochemical solvent by NMP.

Other uses [72] that have been proposed for NMP are as an industrial solvent (spinning of synthetic fibers, surface coatings, solvent welding of plastic films), pigment dispersant (universal tints, pigment incorporation in coating formulations, extension of costly pigments), reaction medium (ethynylation, vinylation), and formulating agent (insecticides, paint removers, surfactant cleaning formulations, deicer for jet fuels, ultraviolet absorber in clear surface coatings). Booklets published by GAF [24,72] are an excellent reference source for a wide variety of applications only briefly alluded to here. Probably the greatest assets of M-PYROL are its powerful solvent action for resins (vinyls, polyesters, polyamides, etc.) and inorganic salts (ferrous, cobalt, zinc, lead, and alkali metal types), besides its low toxicity and volatility). A literature search (CHEMFO #73) is available from GAF [72] describing a variety of uses for M-Pyrol.

2.11 N-VINYL-2-PYRROLIDONE (V-PYROL) (NVP) AND ITS USES

The direct vinylation of 2-pyrrolidone (V-Pyrol) [1,4]* with acetylene under pressure using KOH as catalyst takes place readily at 140 to 160°C and about 15 atm pressure using a nitrogen diluent. The yield of cyclic lactam is 90% or better [3,9,13] using a batch process:

```
CH2 — CH2                                 CH2 — CH2
 |      |                                  |      |
CH2    C=O   + C2H2      KOH              CH2    C=O
   \  /               ---------->            \  /
    N                                          N
    |                                          |
    H                                       CH=CH2
```

The potassium salt of pyrrolidone (2% of the charge) is generally formed in situ by distilling water from the reaction mixture prior to vinylation. The technology here is relatively old and well documented [74]. It is doubtful if any new major innovations have been effected in this reaction.

*GAF registered trademark.

Vinylpyrrolidone, at present, is probably one of the more important Reppe chemicals, since it is sold as monomers, homopolymers, and copolymers. GAF is the sole producer in the United States, while in Europe, BASF is dominant.

V-Pyrol is used in the manufacture of Acryloid lube oil additives, and this market in 1966 was estimated by the author at about 1 million lb of monomer. The lube oils are principally copolymers of NVP with lauryl or octyl methacrylates. An additional 1.5 million lb was probably used in miscellaneous copolymer applications, such as paper coating, paints, cosmetics, and adhesives. Other specialty uses as a pigment dispersant and reaction intermediate have also been suggested [1,24].

Extensive applications data on the copolymerization of V-Pyrol with such monomers as vinyl acetate, styrene, vinyl chloride, acrylonitrile, acrylic acid and esters, maleates, and vinyl ethers is available [24]. The high reactivity of the monomer, together with its polar properties (complete water miscibility, complexing and hydrogen bonding properties), is useful in modifying copolymers through a variation in the hydrophylic-hydrophobic balance of the resulting polymer. The enhanced properties realized by its use in 1 to 20% concentrations are an increase in film strength, dye receptivity, hardness, and adhesiveness. Increased dyeability is particularly useful in vinyl films where printing is often difficult, or in synthetic fibers where superior dyeing and moisture regain are important.

2.12 POLYVINYLPYRROLIDONE (PVP)

The homopolymer has been commercially available for about 22 years in the United States [3,4,24], but longer in Europe. An original use developed in Germany during the war years was its use as a blood plasma expander or blood substitute [4,75,77,78]. The polymer known as *Periston* was a low-molecular-weight (∿40,000), low-viscosity (k = 30) composition that was used as a 20% solution (filtered and sterilized at 120°C) in a specially formulated mixture of salts. Periston was claimed to have saved many thousands of soldiers' lives during the war years, particularly where blood transfusions were not possible. At present, this use for PVP is insignificant compared with its other industrial applications.

Reppe observed that vinylpyrrolidone could be polymerized with ease using 30% hydrogen peroxide in bulk [4]. The resulting exothermic polymerization was allowed to reach 180 to 190°C, whereupon the molten polymer was discharged onto trays, allowed to harden and subsequently ground to a fine white powder.

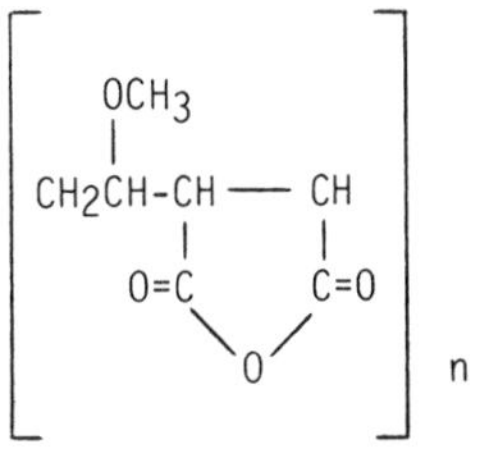

```
┌                      ┐
      OCH3
      |
  CH2CH-CH —— CH
         |     |
       O=C     C=O
          \   /
           O          n
└                      ┘
```

Bulk polymerization was found to be unsatisfactory for commercial production (uncontrollable polymerization, darkening, decomposition) except for the production of the low-viscosity Periston. It was observed [4] that polymerization in aqueous media, using H_2O_2 with either ammonia or amines or their salts, resulted in rapid, easily controlled polymerizations, effected at lower temperatures and with less catalyst. The resulting polymers had much higher viscosities (k values), and were isolated by spray drying techniques. Present-day manufacturing methods are proprietary but are believed to use similar techniques.

In Table 2-5 are summarized applications for various polyvinylpyrrolidone (PVP) polymers. These applications are described in greater detail in Secs. 2.12.1 to 2.12.4.

2.12.1 Uses for PVP

Polyvinylpyrrolidone is gradually decomposed in the human body [4]. Extensive oral, parenteral, topical, and inhalation studies with PVP have shown it to be nontoxic and suitable for both food and drug use [24,77]. The above properties, and its resemblance to natural polyamide or lactam structures, have gained PVP acceptance by the FDA in food and beverage applications, such as the clarification of beer, wine, whiskey, vinegar, and fruit juices.

Polyclar AT [24,81] is the standard PVP grade used in beverage clarification. It is available as a high-molecular-weight, water-and-solvent (especially ethanol-H_2O)-insoluble polymer, in the form of a white powder (95% minimum PVP). It functions through the formation of insoluble complexes with tannins or anthrocyanogens, which, after removal by filtration, prevent haze formation of cold-stored beverages.

Table 2-5 Applications for Polyvinylpyrrolidone (PVP) Polymers

Product	Composition	Applications
Polyvinylpyrrolidone (PVP) [4,24,79-82]		FDA approved for food and drug use; blood plasma expander, nontoxic, biodegradable, varied uses
Polyclar AT [24,81]	Standard PVP grade (white powder)	Beverage clarification aid (fruit juices, wines, beer); chill proofing of beer, complexing agent
Plasdone [73,83,84, 86]	Pharmaceutical grade PVP	Tablet manufacture (Granulating agent); tablet coating; liquid dosages; topical preparations; stabilizer, dispersant, drainage aid (syringes); skin creams, hair sprays, shampoos, general cosmetic use
Kolima adhesive polymers [24] (Kolima 35,55,75)	Modified, clear high solids PVP solutions	Superior glue-line strength, sealing of smooth hard surfaces, grease and chemical resistance, superior adhesion at temperature extremes, superior surface wetting, residual tack and viscosity control applications

Ganex V polymers [92,93,94]	Proprietary PVP polymers	Coatings, detergents, pigment dispersants, plastic additives, textiles, petroleum applications, protective colloids for vinyl latices, improved freeze-thaw stability and scrub resistance
Polectron (P) emulsion [24,92, 94] copolymers	PVP copolymer latices: P-130 (ethyl acrylate); P-230 (2-ethylhexyl-acrylate); P-430,450 (styrene); P825L, 845L (vinyl acetate)	Adhesives: remoistenables, pressure sensitive, heat sealing applications Coatings: leather sizes and finishes, metal primers Paper: Board sizing, pigment binding, pre-coating, heat sealing Textiles: fabric laminates, oil repellent finishes, permanent pressing, polyester sizing

During 1966 it was estimated by the author that Canadian breweries utilized about 300,000 lb of PVP in the chill-proofing of beer. American brewers were then using proteolytic enzymes for beer clarification, but were seriously considering the substitution of PVP at a suitable price. The 1966 price of $1.60/lb was then apparently too high to capture any of the $5 million beer enzyme market. However, with the continued growth of PVP in a variety of areas, lower prices are probable, and its utilization in the rapidly growing wine industry, along with fruit juices and beer, seems quite likely. No estimates are available regarding the size of this market, but it is believed to have multimillion dollar potential.

A pharmaceutical grade of PVP known as Plasdone [78]* is extensively used in tablet manufacture, coating applications, liquid dosage forms, and topical preparations. PVP in tablet manufacture is claimed to permit more economical production speeds (granulating agent) and also gives better quality control and flexibility in the manufacture of effervescent and colored products. In liquid formulations it functions as a stabilizer, dispersant, bodying agent, and a drainage aid for syringes. Topical products such as skin creams, dusting powders, foams, and ophthalmic preparations benefit from PVP via its film-forming and slow-release properties. Miscellaneous pharmaceutical uses were estimated by the author to use about 500,000 lb of PVP in 1966. It is believed this market has grown significantly since them.

A substantial use for PVP [24] is present through its use in cosmetics and toiletries (hair sprays and grooming products, tints, creams, shampoos, eye makeup, face powders, etc.). The U.S. hair spray market alone in 1966 was estimated to use up to 4.5 million lb annually, while the total volume in related cosmetic uses probably amounted to 6 million lb, with a potential dollar value of about $6 million. It is quite possible that this market has grown significantly during the last 10 years, in line with the known growth of the cosmetics industry. The advantages of PVP in hair sprays are better hair management, together with luster and smoothness. In shampoos, creams, and shaving preparations it is claimed to be an emulsion and foam stabilizer, besides having bodying and leveling properties. Its lack of toxicity and synergism in acting as a detoxicant and desensitizer makes it useful in skin products such as creams, deodorants, and face powders.

The versatility of PVP in a wide variety of applications has been described by GAF [24]. These uses include adhesives [79],

*GAF registered trademark.

agricultural chemicals [80], beverage clarification [81], coatings [82], cosmetics and toiletries [83,84], detergents and soaps [85], electrical applications [24], fibers and textiles [24,86], lithography and photography [73], paper [87], pharmaceutical and veterinary products [78,88,89], and polymerizations [24]. A number of these applications include not only PVP but modified PVP (Kolima series) and copolymer types (Polectron emulsions), described below.

About 500,000 lb/yr of PVP were used in the manufacture of Zefran (acrylonitrile type) fiber during 1966. Since acrylic fibers have grown from 352 to 521 million lb in the period 1966-1968 [90], it is quite possible that substantial amounts of PVP are now used in the manufacture of acrylic and modacrylic fibers. The use of PVP in this area is not clear. However, since acrylics are known to have low dye and water affinity, the use of the hydrophylic, easily dyed PVP molecule in solvent spinning may lead to a final fiber with improved properties. PVP has been recommended for increased dye receptivity in hydrophobic fibers [24] and has been patented in mixtures of acrylonitrile polymers [91].

2.12.2 Kolima Adehsive Polymers

This line of modified PVP resins is available commercially [24] as clear solutions of high solids content, readily reducible with a wide variety of organic solvents. They are sold in the following grades: Kolima 35, 55, 75, which in specific formulations have been recommended for both solution and film (air dried) compatibility uses.

They are especially effective in sealing smooth, hard surfaces and show good oil, grease, and chemical resistance, besides being efficient adhesives at temperature extremes. They also contribute unusual surface wetting properties, viscosity control, residual tack, and superior glue-line strength in varied formulations [24]. In 10 to 30% concentrations, they are compatible with such resins as acrylates, epoxies, PVC and copolymers, polyurethanes, and sulfonamides. They have also been FDA approved for use as food packaging adhesives [24].

2.12.3 Ganex V and Miscellaneous Polymers

This proprietary class of PVP-based products, which includes Ganex V [92]* has been modified so that they have good compatibility with mineral oil, organic solvents, and other polymers.

*GAF registered trademark.

The polymers, presumably through copolymerization, have been modified so that they exhibit a variation in hydrophobic-lipophilic properties, together with high surface activity. Their resultant emollient properties make them of particular use in the cosmetic industry. They have also been recommended for uses in coatings, detergents, petroleum applications, pigment dispersants, plastic additives, and textiles. They are available [92] as 100% active products or as 50% isopropanol solutions: Ganex V-216, V-220, V-516, V-804, V-816, and V-904.

A potentially important use for PVP polymers is their role as protective colloids, particularly in vinyl acetate polymerizations [93]. GAF has recommended the following products for polyvinyl acetate lattices:

Polyvinylpyrrolidone: PVP K-90, K-30, K-30/K-90 (blend)
PVP/copolymers (PVP/VA 1 Series)

The polymer latexes are useful for coating, adhesive, and paint formulations. These protective colloids impart to lattices good shelf life, together with excellent mechanical, chemical, and freeze-thaw stability [94]. Improved water and scrub resistance in vinyl acetate films is also claimed.

2.12.4 Polectron Emulsion Copolymers

The versatile film-forming lattices called Polectron emulsion copolymers [24,94]* are available commercially in the following grades, copolymerized with a variety of monomers:

Polectron	130 (ethyl acrylate); 230 (2-ethylhexyl acrylate); 430 and 450 (styrene); 825L and 845L (vinyl acetate)

They are recommended for the following applications:

Adhesives	Remoistenables, pressure sensitives, foam to fabric, heat seals, paper to aluminum
Coatings	Leather sizes and finishes, metal primers
Cosmetics	Hair dyes, acid rinses, wave sets
Household	Aerosol starch, detergent opacifiers
Loose particle binders	Glass fibers, nonwoven fabrics

*GAF registered trademark.

Paper	Board size, pigment binder, precoats, heat seals
Textiles	Fabric laminates, oil repellent finishes, permanent press, polyester size

Polectron films are thermoplastic and remoistenable, adhere well to polar surfaces, machine well, are resistant to oils, and are readily cleaned up with water. A wide variety of applications has been suggested by GAF [94] together with representative formulations.

2.12.5 Technology Details: Overall Reppe (GAF) Process

In 1956 GAF started up a multiproduct production facility at Calvert City, Kentucky based on the ethynylation of aqueous formaldehyde to form propargyl alcohol and butyndiol [76]. The latter product was then converted successively to butenediol, butanediol, butyrolactone, pyrrolidone, methylpyrrolidone, vinylpyrrolidone, and polyvinylpyrrolidone. The GAF facility comprises seven plants, mutually dependent and operating continuously in series, as shown in the flow diagram of Fig. 2-2. Other specialty products which are, or have been, produced are propargyl chloride, bromide, and PVP-iodine complex.

The plant has undergone a number of expansions since its construction, and a sister plant has been in operation since 1968 at Texas City. The acetylene source for the Calvert City operation is calcium carbide acetylene supplied by AIRCO-BOC from its nearby generator facilities, while the Texas City plant uses petrochemical acetylene. Details of process conditions for producing the above Reppe products are covered in Secs. 2.2 to 2.12. The Calvert City facility was the first large Reppe plant built in the United States and was based on German technology and large pilot plant operations at the Grasselli, New Jersey plant of GAF [95]. Du Pont is also practicing Reppe technology for the production of tetrahydrofuran (THF) and butanediol at Houston, Texas (see Secs. 2.7, 2.7.1, 2.8 and 2.17 to 2.17.3).

All Reppe ethynylation plants are designed to handle acetylene safely under pressure and prevent conditions leading to an uncontrollable detonation. Reactors are designed to withstand a pressure rise of 12-fold the initial operating pressure. This pressure rise corresponds to an acetylene deflagration (slow decomposition), and does not result in a detonation where a 200-fold pressure rise is obtained and which is not feasible to contain on a commercial scale. The safe handling of acetylene and the prevention of decompositions and explosions is detailed in Secs. 1.4.7 to 1.4.8.

The following techniques were used for compressing and delivering acetylene to the required process sites [95]:

(a) Low acetylene partial pressures: slow-speed multistage reciprocating compressor submerged in cooling water
(b) High acetylene partial pressures: specially constructed water-sealed centrifugal compressor

A more detailed flow sheet for the production of butyndiol and propargyl alcohol is described in Fig. 2-1. Excess acetylene from the ethynylation reactor is recycled to the vinylation unit where, with fresh acetylene, it is reacted with pyrrolidone to form vinylpyrrolidone, the precursor monomer for the important polyvinylpyrrolidone (PVP) polymer series. A 1975 BASF patent [96] describes a continuous process for the polymerization of vinylpyrrolidone under pressure (1 bar or greater) to PVP polymers having K values of 10 to 35, using a solvent (methanol, THF, xylene, etc.) and peroxide catalyst (dicumyl or di-t-butyl peroxides) at temperatures of 100 to 300 °C and residence times of 3 to 120 min.

2.13 VINYL ETHERS AND THEIR USES

Reppe's original researches [2,3] in the field of vinylation showed that alcohols and acetylene readily interact in the presence of a variety of catalysts to yield vinyl ethers. The best catalysts, however, for vinyl ether formation were alkali metals or hydroxides (1-2% of the alcohol charge). Miller has extensively reviewed [74] the preparation, properties, and technology of a wide variety of vinyl ethers.

$$CH_3OH + C_2H_2 \xrightarrow{KOH} CH_3OCH{=}CH_2$$

Vinylation is generally carried out at 150 to 180°C under pressure using a nitrogen-acetylene mixture (C_2H_2, 100 psig), particularly with lower-boiling alcohols. However, the vinylation method has been extensively studied and modified so that it is possible to operate at atmospheric pressure using either a liquid- [98] or vapor-phase process [99]. High-boiling polar solvents have also been used. In general, low-molecular-weight alcohols react more readily, although it is possible by varying conditions to react longer chain primary and secondary alcohols.

Trans-etherification is also a potentially useful tool for the preparation of higher boiling vinyl ethers from relatively expensive alcohols [100]. The interchange reaction of vinyl isobutyl ether with heat-sensitive higher alcohols such as geraniol-nerol or phenyl ethanol in the presence of a mercuric acetate-amine catalyst is effected in high yield by fractionally distilling off the more volatile alcohol [101].

It would also appear that Reppe's original method [2,4,97] of reacting vinyl chloride with the alkoxides of alcohols under pressure could be of present-day use in the manufacture of specialty vinyl ethers. The cheapness of vinyl chloride (15¢/lb, 1979) and the possibility of using either sodium hydroxide or sodium for alkoxide formation would indicate a practical route for less reactive and more expensive alcohols. The use of polar media (DMF, DMSO) might also further activate this type of reaction.

ICI [102,103] has developed new technology for vinyl ethers which effects the reaction of alcohols with ethylene in the presence of a $PdCl_2$-$CuCl_2$ catalyst using oxygen at about 50°C. This process is closely related to the ethylene-based vinyl acetate process using acetic acid or acetaldehyde with oxygen (cf. Sec. 1.6.11):

$$ROH + CH_2{=}CH_2 + \tfrac{1}{2}O_2 \xrightarrow[CuCl_2]{PdCl_2} CH_2{=}CHOR + H_2O$$

It is claimed the method at sufficient volume can produce methyl vinyl ether at about half its present price. Ethylene-based $PdCl_2$ processes for both vinyl ethers and vinyl esters appear to be favored routes over acetylene in the years ahead.

In Table 2-6 are summarized the more important uses for the vinyl ethers and their copolymers. These applications are described in greater detail in Secs. 2.13.1 to 2.13.4.

The commercial utilization of vinyl ethers in the United States has been relatively slow in spite of wide publicity and substantial development work over the years by companies such as Air Reduction and General Aniline and Film (GAF). It is only in recent years that significant volume uses for vinyl ethers, through

Table 2-6 Application for Vinyl Ethers and Polymers

Product	Composition	Applications
Methyl vinyl Ether [104,106,108]	$CH_3O\text{-}CH{=}CH_2$	Specialty monomer; intermediate; Gantrez resins
Ethyl, n-butyl, isobutyl, decyl vinyl ethers [104, 105-107]	CH_2H_2 + alcohol $\longrightarrow$ $RO\text{-}CH{=}CH_2$	Specialty monomers; copolymers (adhesives) lacquers, paints, plasticizers, pour-point depressants; reaction intermediates
Gantrez M [108-111]	Polyvinyl methyl ether Grades: M-55, M-154 (50% H_2O) M-574 (70% solids in toluene).	Adhesive formulations with high wet tack; adhesion to plastic and metal surfaces; uses: adhesives, coatings, printing inks, textiles
Gantrez AN [112-113]	Copolymer of methyl vinyl ether and maleic anhydride Grades: AN-139, AN-149, AN-169.	Polyelectrolyte in aqueous media; uses: hair sprays, shampoos, detergents, leather, paper coating, pigment grinding, nonwoven fabrics, topical remedies, tablet coating, photoreproduction
Gantrez VC [110,114]	Copolymers of alkyl vinyl ethers (e.g., isobutyl vinyl ether) and vinyl chloride.	Traffic, marine and maintenance surface coatings, foundation sealants, corrosion resistant paints, flame retardancy for vinyl substrates, primer coat for metals, paper coating

copolymer applications (Gantrez resins), have been realized in the United States by GAF.

From Reppe's writings it is apparent he regarded vinyl ethers quite highly, particularly in terms of their industrial potential. Based on extensive polymerization and application studies, the Germans were successful in developing commercial uses in areas such as adhesives, lacquers, paints, plasticizers, pour-point depressants, and textile aids. A series of homo- and copolymers of vinyl ethers known as *igevins*, *oppanols*, and *densodrins* were illustrative of this effort.

During the period 1960-1970 the General Aniline and Film Company, through its Antara Chemicals Division, successfully introduced vinyl ether technology to the United States marketplace through excellent trade literature, publications, and commercial development technical seminars [104-109]. Both short-chain (methyl, ethyl, n-butyl, and isobutyl) and long-chain (isoctyl, decyl, dodecyl, hexadecyl, and stearyl) vinyl ethers were and are offered commercially [109], together with polymer applications information [105-109]. The vinyl ethers are reactive molecules and are potentially useful intermediates for a wide variety of new products [104,105].

There is no published information regarding the production or sale of vinyl ethers or their polymer derivatives. However, according to the author's estimate during 1964, the total production of vinyl ethers was about 2 million lb. However, since then there is believed to have been rapid growth, particularly in the Gantrez M, AN, and VC lines of methyl vinyl ether polymers. The announced expansion [110] of the GAF copolymer line, since their commercialization [111] in 1966, testifies to this rapid growth.

2.13.1 Gantrez Polymers: Gantrez M

This line of vinyl ether-based polymers is comprised mainly of the following: Gantrez M (Polyvinyl methyl ether [108], Gantrez AN (copolymer of vinyl methyl ether-maleic anhydride), and Gantrez VC (copolymer of alkyl vinyl ethers and vinyl chloride).

Gantrez M, a tacky semisolid, is composed of methyl vinyl ether units in a linear distribution. The polymer is sold commercially [108] in either water (50% solids, grades M-154 and M-155) or toluene (70% solids, M-574 and 50% solids, M-555). Gantrez M-154 and M-574 are relatively low molecular weight grades with k values of 40, respectively. Higher molecular weight polymers (k values 53) are typified by M-155 and M-555, which have greater cohesive strength but less tack than the lower molecular weight analogs.

The attractive properties of these polymers include their high wet tack in adhesive formulations, excellent adhesion to plastic and metallic surfaces, broad compatibility with both water and solvent soluble polymers, excellent solubility in a wide variety of solvents and water, and the ability to function as nonmigratory plasticizers. They have been recommended for use in adhesives, coatings, printing inks, heat sensitization uses (paper coating, textile finishes, nonwovens, rug backing), polymerization (particle size control), latex paints, and textiles.

About 500,000 lb of Gantrez M was used as a heat stabilizer for dip coating of rubber goods during 1966-67. This and other suggested uses in nonwovens, textiles, and rug backing are believed to have grown substantially since then.

2.13.2 Gantrez AN [112]

This line of polymers, based mainly on the copolymerization of maleic anhydride and methyl vinyl ether (MVE), are probably the principal factors in the rapid growth of vinyl ether-based polymers. The combined structural features of high polarity (methyl vinyl ether) and reactive functionality (anhydride units) make for a polymer molecule with not only interesting properties, but wide applicability via its reactivity potential (alcohols, amines, nonionic surfactants).

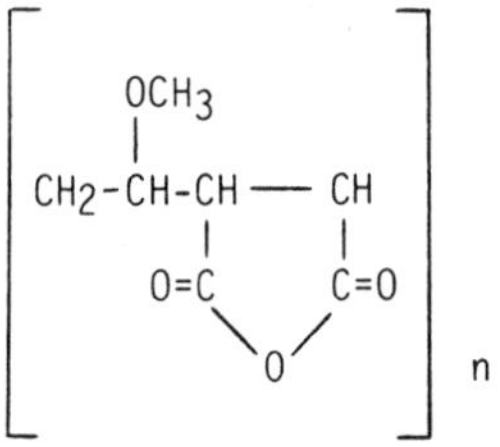

```
[                              ]
      OCH3
      |
 CH2-CH-CH —— CH
         |     |
       O=C     C=O
          \   /
           O                   ]n
```

The polymer is available in a variety of grades: Gantrez AN-119 (low molecular weight), AN-139 and -149 (medium molecular weight), and AN-169 (high molecular weight). The polymer can be regarded as a polyelectrolyte since it is soluble in water over the entire pH range. It shows excellent compatibility with water soluble gums, resins, plasticizers and metallic salts. Its dispersing, stabilizing, and protective colloid properties make it an effective thickener and surfactant, useful in either aqueous or mixed solvent-water systems. Unlike the parent vinyl ethers, the homo- and copolymers show excellent acid stability. The polar nature of the polymer is reflected in films that have good adhesion to a variety of surfaces. The anhydride site is also an interesting handle for the formation of metallo-organic polymers in aqueous or solvent systems to yield linear or ladder-type structures.

GAF has summarized a wide area of applications [111], which include adhesives [112], coatings, polymerizations, detergents, hair sprays, shampoos, leather, paper coating, topical remedies, tablet coating, photoreproduction, pigment grinding, nonwoven fabrics, synthetic fabric sizes and finishes, besides numerous smaller uses. The polymer's low toxicity (oral and skin), and FDA approval in such uses as food packaging, pharmaceuticals, and cosmetics, makes it quite versatile.

Gantrez AN-8194 [104,105] is an analog that utilizes octadecyl vinyl ether with maleic anhydride in place of methylvinyl ether. The copolymer is offered in toluene as a release coating for pressure-sensitive adhesives, as an antiblocking agent, and as a film-former for aerosol products. Other Gantrez analogs [104] available for a variety of specialty applications include AN-3152, AN-3952, Thickener L, AN-3159, and Appretan D and N.

Trade rumors during 1967 indicated that demand for the Gantrez AN series compelled GAF to allocate the product. Its volume use at that time was estimated by the author to be as high as 5 million lb/yr. A major use then, and also now, is as a hair spray additive. Whether this use has replaced some of the growth of polyvinylpyrrolidone in this area is not known. Uses (1967) in liquid detergents and textiles were about 500,000 lb. Since the production of vinyl ether polymers dates to only 1966, and since the GAF production facilities have expanded, strong growth in the above and related areas appears quite likely.

2.13.3 Gantrez VC [114]

This class of copolymers is based on alkyl vinyl ethers and vinyl chloride [109]. The isobutyl vinyl ether (IBVE) copolymer is recommended for surface coating applications (solution polymers) such as marine and maintenance use. Gantrez VC is claimed to have outstanding corrosion resistance and is a flame retardant on vinyl substrates. Its films (paint, adhesives, sealants, etc.) are particularly useful where moisture and abrasion resistance are needed (traffic and swimming pool paints, foundation sealants). The excellent adhesion of such films to base metals and prime coats, together with their flexibility and compatibility with many solvents and resins, makes then quite versatile.

The copolymerization of methyl vinyl ether with such monomers as acrylonitrile, vinyl acetate, vinyl chloride, and vinylidene chloride in a low-pressure system using free radical initiators has been described by GAF [106], together with properties and applications [109]. The Gantrez VC types are probably the most important, commercially, of the new vinyl ether copolymer types being offered in the United States.

BASF has developed a vinyl chloride-isobutyl vinyl ether (3VC:1 IBVE) analog of Gantrez VC known as Vinoflex MP-400. Approximately 4 to 5 million lb were sold in Europe in the period 1966-1967, with an initial penetration of the U.S. market estimated at about 0.5 million lb. With both GAF and BASF offering the product, commercially increased growth since 1967 is probable.

2.13.4 Koresin

Reppe [3] observed that the vinylation of phenols with zinc or cadmium naphthenates as catalysts resulted in the formation of ortho hydroxystyrenes, which polymerized in situ to resinous products. He showed that the reaction involved rearrangement (Claison) of the initially formed vinyl ether:

$$HOC_6H_4C(CH_3)_3 + C_2H_2 \xrightarrow{180\text{-}200^\circ C.} [CH_2{=}CHOC_6H_4C(CH_3)_3] \rightarrow HO(CH_2{=}CH)C_6H_3C(CH_3)_3 \rightarrow \text{Resins}$$

Koresin

Reppe considered phenol-acetylene polymers to have particularly attractive properties and predicted a great future for them. Koresin, in particular, besides having a high degree of tolerance for both buna and natural rubber, improved compounding, adhesion, and extruding properties of rubbers in general.

Large-volume uses for Koresin never materialized, particularly in the United States. Limited amounts (probably 1-2 million lb) were sold for use as a tackifier for natural rubber. The product was manufactured by GAF and was later discontinued. It has, however, been listed in their product line.

In view of a variety of commercially available mono- and polyphenols, it would appear that acetylene-phenol polymers, particularly copolymer types, should enjoy more important commercial uses than they do at present. Perhaps the key to marketing these materials lies in handling them in suitable solvent systems and proprietary mixtures rather than as the solid polymers. Their properties indicate uses not only in rubber technology but also in adhesives and surface coatings [115].

2.14 VINYL FLUORIDE (VF) AND 1,1-DIFLUOROETHANE (DFE)

During the period 1947-1950 a substantial amount of industrial research was reported in the patent literature on the preparation of vinyl fluoride (VF) and ethylidene fluoride (1,1-difluoroethane, DFE).

$$\underset{}{CH{\equiv}CH} + HF \longrightarrow \underset{VF}{CH_2{=}CHF} \xrightarrow{HF} \underset{DFE}{CH_3CHF_2}$$

Miller (74) has provided an excellent review of much of the patent art leading to the commercialization of vinyl fluoride. Mercury compounds deposited on carbon have been one of the preferred catalyst systems [116]. The formation of vinyl fluoride is accompanied by varying amounts of difluoroethane, and the reaction can be directed to high yields of the latter if desired [117].

The separation of acetylene from vinyl fluoride presents a process problem, since both compounds have fairly close volatilities. Although methods such as reaction with HCl or the use of solvents have been proposed [74], it is believed that the preferred industrial process for VF involves cracking DFE:

$$CH_3CHF_2 \xrightarrow{\Delta} CH_2{=}CHF + HF$$

Pyrolysis in the range of 200 to 400°C using catalysts such as alumina or calcium sulfates has been reported [118].

The principal U.S. producers of VF are Du Pont (Louisville) and Diamond Shamrock (Houston). The principal volume use for VF is the manufacture of polyvinyl fluoride (PVF) [119,120] for exterior paints on surfaces such as hardboard and aluminum siding, steel building panels, asbestos-impregnated felt roofing, and fiber-glass paneling. Du Pont introduced PVF in 1963 under the registered trademark Tedlar and found early market exploitation of this material slow. However, Du Pont has reported [119] that PVF sales in 1968 were 50% higher than in 1967.

At present, PVF is involved in a key economic battle with acrylic polymers in the large hardboard market [119]. Rohm and Haas introduced a competitive acrylic known as Korad in 1967 that has established a position in outdoor surface coating of highway signs, decking, ABS, and PVC sheeting. To further strengthen its basic position in acrylics, Rohm and Haas has announced significant expansion of its acrylic capacity at Deer Park, Texas to a rated volume of 400 million lb/yr [121].

Since lumber prices and construction costs have mounted rapidly in recent years, and since plywood production lacks long-range flexibility for continuous production, the trend in recent years has been to the cheaper hardboard fibrous wood products which can be geared to production needs, and utilize both wood and related materials. Du Pont's Tedlar is at present being used on hardboard (U.S. Plywood) at a thickness of 1.5 mils [119]. Erosion tests have shown that 75% and 70% of the original PVF film should remain after 25 and 30 years, respectively. Rohm and Haas claims similar advantages for Korad in weatherability. A PVF-pigmented product used as a 1.5-mil film gives a cost of 3 to 3.5¢/ ft^2 of board. At present, a 3-mil acrylic film is believed to be competitive with PVF. Rohm and Haas is reported to be developing a 2-mil film which would, in turn, drop costs to the 2.5¢ range to aid in this marketing battle.

The total amount of acetylene consumed [122] in 1966 for the manufacture of vinyl fluoride and the related monomer, vinylidene fluoride, has been estimated at about 1 million lb. In view of the 50% increase in PVF sales in the period 1967-1968 (Du Pont) it must be assumed that steady growth in this specialty polymer area is still present.

2.14.1 Vinylidene Fluoride

The monomer is produced by the chlorination of difluoroethane followed by the cracking of the intermediate chlorodifluoroethane [120,123]:

$$CH_3CHF_2 + Cl_2 \longrightarrow CH_3CClF_2 \xrightarrow{\Delta} CH_2{=}CF_2 + HCl$$

The principal U.S. producers of the monomer are Allied Chemical, Du Pont, and Penn Salt. Its commercial use appears to be directed almost exclusively to the manufacture of homo- and copolymers. In 1962 Penn Salt constructed a semiworks facility of about 200,000-lb capacity and by 1968 had completed a commercial plant for both the monomer and polymer.

Polyvinylidene fluoride is a tough linear polymer, very resistant to exterior weathering, sunlight, and chemical attack by acids, alkalies, oxidizers, halogens, and many organic compounds. Highly polar solvents (dimethylacetamide, γ-butyrolactone) and polar ketones will, however, cause swelling and will yield colloidal suspensions, which have been used in coating formulations.

The greatest potential for the homopolymer appears to lie in its uses as an exterior coating for steel, aluminum, wood, etc., where it will be in competition with PVF [124] and acrylics. However, due to its excellent mechanical properties (strength, resistance to impact, abrasion, deformation under load, fatigue) and chemical resistance, it is used in diverse specialty applications, which include coatings and linings for pipe, tanks and process equipment. In the aerospace field, the need for high reliability and instant response to components, besides chemical inertness, is reflected in the use of polyvinylidene fluoride for expulsion bladders, valve seats, lip seats, wires and cables, and frangible diaphragms.

The present total market for the homopolymer and its copolymers (ethylene, tetrafluoroethylene, and chlorotrifluoroethylene) is not known. However, since 1966 about 1 million lb of acetylene was used between VF and vinylidene fluoride, and it is likely that the growth of the latter monomer parallels that of VF.

2.14.2 Miscellaneous Specialty Vinyl Monomers

Miller has provided an excellent review of the synthesis of a wide variety of vinyl compounds, including ethers, esters, and nitrogen types (Ref. 74, pp. 304-361). During the past 30 years a large amount of research and development work has been expended in industrial laboratories both in the United States (Air Reduction, GAF, Dow, etc.) and in Europe (BASF, British Oxygen, ICI) on vinyl derivatives. Some of the more commonly studied types and examples were:

Vinyl esters: Oleic, stearic, pivalic, adipic, benzoic acids.
Vinyl ethers: Long-chain aliphatic alcohols, branched oxo-alcohols, allyl alcohols, glycols, phenols.
Vinyl nitrogen types: Amides, carbazole, caprolactam, piperidine, amides, pyrrolidone analogs.

A substantial amount of marketing and commercial development time was also spent attempting to develop markets for such monomers. To date, no significant reported markets have developed for such specialty monomers, although commercial potential is still believed present. As with many specialty products, an induction period of up to 20 years is possible. For example, vinylcarbazole has been much studied [4], and its homopolymer is considered to have unique electrical and thermal properties. It has been considered in dielectric applications and in electrophotoreproduction processes, where it is a potential replacement for zinc oxide (Ref. 74, pp. 304-361). However, to date no significant U.S.

market has yet developed. Continued interest in vinyl substitutes (hydantoins, oxazolidones) for N-vinylpyrrolidone is still believed present due to the good growth in PVP applications. Also, there is potential for vinylated heterocyclic structures as specialty monomers for use in latex emulsions to increase adhesion to exterior surfaces exposed to excess moisture and weathering conditions.

2.15 SECONDARY AND TERTIARY ACETYLENIC ALCOHOLS AND GLYCOLS

2.15.1 Alkynols

The Reppe copper acetylide system, while being an excellent production route for the ethynylation of formaldehyde to propargyl alcohol and butyndiol (see Secs. 2.1-2.4), cannot be used effectively for other aldehydes and ketones due to low conversions and yields. The preferred commercial routes for alkynols involve either the use of sodium acetylide or potassium hydroxide (Favorsky method). A catalytic method using basic resins is also possible for the formation of tertiary alkynols, but is limited to low-molecular-weight polar ketones (see Ethynylation Technology Sec. 2.15.6):

$$\text{R-CHO} + \text{NaC}_2\text{H} \xrightarrow{0\text{-}50^\circ} \left[\text{R-}\underset{\displaystyle\text{OH}}{\underset{|}{\text{C}}}\text{HC}\equiv\text{CH}\right]\text{Na} \xrightarrow{\text{H}_2\text{O}} \text{R-}\underset{\displaystyle\text{OH}}{\underset{|}{\text{C}}}\text{HC}\equiv\text{CH}$$

$$+ \text{NaOH}$$

$$\text{R}_1\text{R}_2\text{C=O} + \text{CH}\equiv\text{CH} + \text{KOH} \xrightarrow{0\text{-}30^\circ} \left[\text{R}_1\text{R}_2\underset{\displaystyle\text{OH}}{\underset{|}{\text{C}}}\text{-C}\equiv\text{CH}\right]\text{KOH}$$

$$\xrightarrow{\text{H}_2\text{O}} \text{R}_1\text{R}_2\underset{\displaystyle\text{OH}}{\underset{|}{\text{C}}}\text{-C}\equiv\text{CH} + \text{KOH}$$

Both methods have been extensively documented in earlier [125, 126] and more recent [127,128] texts on acetylene chemistry. Both techniques have also been utilized with a wide variety of aldehydes and ketones. The use of polar solvents such as ethers, acetals, amides, amines, and nitriles has been observed to have an activating effect upon the reaction [10,127,129].

The Favorsky method, however, cannot be used with formaldehyde, since Cannizzaro and aldol side reactions completely overshadow ethynylation. Acetaldehyde, also easily aldolized, however, can be converted to 1-butyn-3-ol in either liquid ammonia [10,127] or polar organic solvents [127,130] when NaOH or KOH are employed. This method supplements the well-known alkali metal (Li, Na, K) acetylide route. The Reppe route, while it can be employed for acetaldehyde, is not practical industrially, since long reaction times (22 hr) must be employed at 125°C to obtain an 18% conversion to 1-butyn-3-ol (B) and a 10% conversion to 3-hexyne-2,5-diol (HD) [3,5].

$$CH_3CHO + C_2H_2 \xrightarrow{CuC_2H} \underset{(B)}{CH_3\underset{OH}{\underset{|}{C}}HC{\equiv}CH} + \underset{(HD)}{CH_3\underset{OH}{\underset{|}{C}}H\text{-}C{\equiv}C\text{-}\underset{OH}{\underset{|}{C}}HCH_3}$$

Reppe observed that formaldehyde in a 12-hr period at 90 to 100°C gave conversions to butyndiol up to 95% [3,5]. The Favorsky route, or the use of alkali metal acetylides with acetaldehyde, gives 50 to 75% conversions to (B) in less than 1 hr at 0 to 20°C.

2.15.2 Alkyndiols

Acetylenic glycols (secondary and tertiary) are best formed by the use of powdered potassium hydroxide, which is also the preferred catalyst for alkynols [127]:

$$2R_1R_2C{=}O + CH{\equiv}CH + KOH$$

$$\xrightarrow[20\text{-}50°]{} \left[R_1R_2\text{-}\underset{OH}{\underset{|}{C}}\text{-}C{\equiv}C\text{-}\underset{OH}{\underset{|}{C}}\text{-}R_1R_2\right] KOH$$

$$\xrightarrow{H_2O} R_1R_2\text{-}\underset{OH}{\underset{|}{C}}\text{-}C{\equiv}C\text{-}\underset{OH}{\underset{|}{C}}\text{-}R_1R_2 + KOH$$

Glycol formation is best carried out at 20 to 50°C, while alkynol formation predominates below 10°C. It is readily possible by variations in temperature, reaction stoichiometry, and feed rates to direct the reaction predominantly (90%) to either the alkynol or alkyndiol. Methyl ketones, even with chains greater than 10 carbons, react well (conversions 70%) to yield tertiary diols. Both straight-chain and branched aldehydes (exception: formaldehyde and acetaldehyde) react reasonably well (conversions 50-75%) to yield secondary acetylenic diols [125,20]. Aldolization can be minimized, but not completely eliminated, when using powdered KOH.

2.15.3 Alkali Metal Hydroxide Complexes and Disproportionation of Alkynol Base Complexes

Tertiary alkynol and alkyndiol formation with alkali metal hydroxides has been shown to involve the intermediate formation of relatively stable complexes between the base and the acetylenic hydroxy compound, as shown in the above equations (Sec. 2.15.2). Tedeschi and coworkers (Ref. 10, p. 2480) have prepared and isolated these complexes via the reaction of powdered alkali metal hydroxides with either tertiary alkynols or alkyndiols, respectively, or via the reaction of acetylene, KOH, and acetone. The KOH complexes of 3-methyl-1-butyn-3-ol and 2,5-dimethyl-3-hexyn-2,5-diol were also shown to be capable of functioning as catalysts for the formation of methyl butynol [131,132] (see Sec. 2.15.6). Rutledge [127] and Viehe [128] also have summarized pertinent details of this work as it affects the general concept of ethynylation.

Precursor species leading to the formation of the alkynol or alkyndiol-base complexes were also shown to involve acetylene-base complexes [127]:

$$C_2H_2 + KOH \longrightarrow C_2H_2(KOH)$$

These relatively unstable complexes readily decompose to acetylene and the alkali metal hydroxide on heating above 70°C. They are readily formed in either liquid ammonia [10], organic solvents [10], or in excess liquid acetylene under pressure [133]. The alkynol-NaOH complexes do not form to a significant extent in moderately polar organic solvents (dioxane, methylal), but they can be readily formed in liquid ammonia [131,132] in yields of 85% using 3-methyl-1-butyn-3-ol. The analogous reaction with KOH is quantitative in either liquid ammonia or organic solvents. Lithium hydroxide, in contrast to the well-known activity of lithium acetylide for alkynol formation, is completely ineffective as a catalyst for ethynylation. Although it readily forms a $LiOH(C_2H_2)$ complex [133], it fails to yield the corresponding alkynol or alkyndiol-LiOH adduct [131] even in liquid ammonia. Sodium hydroxide, although ineffective in acetal or ether solvents, can be used effectively in liquid ammonia [131], confirming the formation of alkynol-NaOH complexes.

An unusual solid-phase transformation (disproportionation) of the KOH complex of 3-methyl-1-butyn-3-ol into the KOH adduct of the corresponding acetylenic diol, 2,5-dimethyl-3-hexyn-2,5-diol was observed [10] to take place via the following route:

$$2\ |\ (CH_3)_2C(OH){-}C{\equiv}CH\ |\ KOH \longrightarrow |\ (CH_3)_2C(OH){-}C{\equiv}C{-}C(OH)(CH_3)_2\ |\ KOH$$

MB(KOH) DH(KOH)

$$+\ KOH(C_2H_2) \rightleftharpoons KOH + C_2H_2$$

The complete disproportionation takes place gradually at room temperature and more rapidly below the decomposition or cleavage temperature (50-70°C) of the complex. No change in appearance (darkening or evidence of acetone condensation products) is noted in the white, finely divided solid complex during the transformation. This observation means that free acetone is not present during the disproportionation process at 25 to 30°C. During a 30-month period at room temperature, complete conversion of the alkynol adduct (MB-KOH) into the alkyndiol complex (DH-KOH) was noted, while at 35 to 40°C a 14% conversion to the diol complex is realized in 6 hr. The NaOH-methyl butynol complex undergoes a similar transformation but at a faster rate [9,132].

2.15.4 Mechanism of Acetylenic Diol Formation

Tedeschi [132] has proposed the following mechanism for the stoichiometric formation of tertiary acetylenic glycols based on a rapid attack of ketone upon the acetylenic-KOH complex. This heterogeneous reaction is viewed as involving chemisoption (via hydrogen bonding) of the ketone on the surface of the complex, followed by rearrangement of the intermediate complex to the stable diol-base adduct:

$$(CH_3)_2\text{-C=O---H-C}\equiv\text{C-H---O=C-}(CH_3)_2 \longrightarrow$$

KOH

$$\text{(a)}\ (CH_3)_2\text{-C---C} \equiv \text{C---C-}(CH_3)_2 \longrightarrow$$

O---H KOH H---O

$$\text{(b)}\ (CH_3)_2\text{-C-C} \equiv \text{C-C-}(CH_3)_2$$

OH---KOH---OH

diol-KOH adduct

The failure of the KOH complex of 3-methyl-1-butyn-3-ol to react with acetone in ether solvents (30-40°C) was cited as evidence for the direct attack of acetone upon the KOH-C_2H_2 complex to yield the diol complex:

$$\left[(CH_3)_2\text{-C(OH)-C}\equiv\text{CH}\right]\text{KOH} + (CH_3)_2\text{C=O} \longrightarrow \text{no reaction (diol-KOH adduct)}$$

When diol formation is carried out in the recommended range of 30 to 40°C, a stoichiometric (2 m ketone, 1 m KOH, 1 m C_2H_2) ratio of reactants is not required. Usual methods in the literature [127,128] cite saturating a KOH-solvent slurry with excess acetylene, followed by addition of ketone. Since this method is known to give high conversions (95%) to the alkynol-KOH complex at 0 to 5°C, the high conversions (85-95%) of acetylenic diol observed in the 30 to 40°C range strongly supports a mechanism in which the intermediate formation of alkynol is not required.

Base-catalyzed cleavage of alkynols back to their starting components, acetylene and ketone, becomes significant above 50°C, and provides another route to acetylenic diol formation. Acetylenic diols also undergo the cleavage reaction, yielding first the alkynol and the starting carbonyl compound. The alkynol on further heating will reverse to acetylene and an additional molecule of carbonyl compound. The base-catalyzed cleavage of secondary and tertiary acetylenic diols can be used as a preparative method for acetylenic alcohols but is not competitive economically with the primary route from the carbonyl compound and acetylene. The base cleavage route is also characterized by the formation of polyaldol byproducts derived from the respective aldehyde or ketone used to prepare the acetylenic hydroxy compound.

The base-catalyzed cleavage of alkynols, with the resultant formation of acetylenic diols, is shown below and can be defined as a second-stage process to differentiate it from the first-stage (primary) route based on carbonyl compound and acetylene.

$$2(CH_3)_2\text{-}\underset{\text{OH}\cdots\text{KOH}}{C}\text{-}C\equiv CH \xrightarrow{\Delta} 2(CH_3)_2C\text{-}O + \underset{\text{KOH}}{CH\equiv CH} + C_2H_2\uparrow$$

$$\longrightarrow (CH_3)_2\text{-}C\text{-}C \equiv C\text{-}C\text{-}(CH_3)_2$$
$$\text{OH}\cdot\cdot\text{KOH}\cdot\cdot\text{OH}$$

This second-stage conversion to alkyndiol is often confused with the primary, or first stage, conversion of ketone and acetylene to diol, particularly at higher temperatures (40-60°C) where both type of reactions can proceed together.

A third-stage conversion to alkyndiol involves the disproportionation of the alkynol-KOH complex as discussed in Sec. 2.15.3. This route to diol, however, is not important unless lower temperature (25-30°C) conditions over a long period of time are desired for selective formation of alkyndiols. The disproportionation (third stage) route is the slowest for acetylenic diol formation, with the primary (C_2H_2-KOH-ketone) route being the most rapid.

2.15.5 Acetylene Solvents

Solvents that can hydrogen bond with, and consequently dissolve, large amounts of acetylene are particularly useful in separating it from impurities [134]. Typical of good acetylene solvents are polar, low-molecular-weight ethers, acetals, amines, nitriles, and ketones. Outstanding solvents are typified by the aprotic solvents, dimethyl sulfoxide (DMSO), dimethyl formamide (DMF), dimethyl acetamide, hexamethyl phosphoramide, and N-methyl pyrrolidone [134]. Probably in a class by itself in overall qualities of high acetylene solubility, activating or cocatalyst properties, ease of recovery, ease of acetylenic product isolation, and cheapness, stands liquid ammonia. For example, thousands of gallons of liquid ammonia are recovered and recycled daily in the various ethynylation processes involved in the commercial manufacture of vitamins A and E and related perfumeries. Liquid ammonia ethynylation technology is currently utilized by Hoffmann-La Roche (Nutley, N.J.) and Chas. Pfizer (Groton, Connecticut).

A useful acetylene solvent must be capable of bonding hydrogen more strongly with acetylene than with itself. Consequently, water and alcohols are in most cases inferior solvents, due to strong intermolecular attraction [134]:

$$NH_3 + HC{\equiv}CH \rightleftharpoons H\text{-}C{\equiv}CH\cdots NH_3 + H_3N\cdots HC{\equiv}CH\cdots NH_3$$

$$(H_2O), \quad H\text{-}O\text{-}H\cdots\cdots O(H)(H)\cdots\cdots H\text{-}O\text{-}H$$

Although water will form solid hydrates [133,135] with acetylene under pressure (analogous to CO_2), such adducts fail to solubilize acetylene more than polar solvents.

McKinnis [134] has developed a useful formula for determining superior acetylene solvents based on the hydrogen bonding power of various donor centers:

$$s = kN^{\frac{1}{2}}\,(\chi_A - \chi_B)d^3$$

S equals grams of acetylene dissolved per mole of solvent (25°C and 1 atm), while $N^{\frac{1}{2}}\,(\chi_A - \chi_B)d^3$ defines a donor center and its relative hydrogen bonding ability based on Pauling's electronegativity (χ_A or χ_B) values. The method was shown to have a predicting accuracy of within 10% as checked by experimental data and was used to predict the outstanding acetylene affinity of hexamethyl phosphoramide $[(CH_3)_2N]_3PO$ before its synthesis.

Miller [136] has extensively reviewed solubility parameters of acetylene with a wide variety of solvents at both atmospheric and elevated pressures. Tedeschi et al. have reported [133] the interaction of liquified acetylene with stoichiometric and excess amounts of both polar and nonpolar materials under pressure. At 40°C, five degrees above the critical temperature (35°C) of acetylene, hydrogen bonding was still significant with 80 to 94% of the acetylene in the liquid phase. Besides ammonia, methylamines and N-methylpyrrolidone were excellent solvents at 0 to 40°C under pressure.

Polar solvents such as butyrolactone, N-methylpyrrolidone, dimethylformamide, and NH_3 are used to extract acetylene from pyrolysis process streams.

2.15.6 Ethynylation Technology

Liquid Ammonia-Alkali Hydroxides Route The ethynylation of ketones and aldehydes can be carried out with high efficiency in liquid ammonia using catalytic amounts of either sodium or potassium hydroxide [129,137]. It is possible to convert 6 to 18 mol of a ketone or aldehyde in approximately 500 ml of liquid ammonia, using 1.5 mol of base, to the corresponding alkynol in conversions of 50 to 100%.

Today, the alkali hydroxide-liquid ammonia system probably represents one of the more versatile and productive systems for the preparation of a variety of secondary and tertiary alkynols. It can produce considerably more alkynol from ketones such as acetone or cyclohexanone than the corresponding organic solvent route. Most aldehydes, with the exception of formaldehyde, react well, but optimum yields and catalytic conversions are obtained with aliphatic methyl ketones. The catalytic effect (conversion) is defined as the moles of alkynol produced divided by the moles of base used times 100. The results given in Table 2-7 illustrate the versatility and production potential of this method [10,129,131].

The process is operated at modest pressures (100-350 psig) and temperatures (-30 to 40°C) using a 1.3 to 4.0-fold excess of acetylene over theory. Optimum catalytic results are obtained at 20 to 40°C, using at least a 1.5 to 2.0-fold excess of acetylene. At high concentrations (loading) of carbonyl compound (18-24 m/ 500 cm^3 liquid ammonia) a drop in conversion from the 85 to 95% level to the 50 to 75% range is customary, as the activating effect of liquid ammonia is diluted. Reactive ketones such as acetone and cyclohexanone, however, still give excellent (75 and 71%, respectively) conversions to distilled product at the 18 m level (see Table 2-7).

Vinyl and isopropenyl acetylenes [10,129] react as well as acetylene in the catalytic ammonia system to yield the corresponding eneynols. The reaction of acetone (6 m) with isopropenyl acetylene (IPA, 6.0 m) in 500 ml of liquid ammonia with 1.5 m potassium hydroxide yields 2,5-dimethylhex-5-ene-3-yn-2-ol in 69% distilled conversion.

The alkali hydroxide-liquid ammonia route also has versatility for producing acetylenic diols [9,10,129,138] by varying typical reaction parameters such as feed ratios, temperature, and pressure. The author has shown that alkali metal hydroxide complexes (Na, K, Cs, Rb) of either 3-methyl-1-butyn-3-ol (MB) or 2,5-dimethyl-3-hexyn-2,5-diol (DH) can be used in place of alkali metal hydroxides to give equal results. This data substantiates the mechanistic role of these adducts in catalytic ethynylation.

A variation of the ammonia system has been reported by Moore and Tedeschi [139] in which excess liquid acetylene containing a catalytic amount of ammonia and KOH was used to ethynylate

Table 2-7 Catalytic Formation of Alkynols in Liquid Ammonia[a]

Carbonyl Compound	2° or 3°[b] Alkynol	Moles Carbonyl Compound	Percent Distilled Conversion Based On >C=O	KOH
$(CH_3)_2C{=}O$	MB	6-12	95-82	381-660
	MB	18	75	902
	MB	24	52	835
$C_2H_5COCH_3$	MP	18	67	800
$(CH_3)_2CHCH_2COCH_3$	DMH	17	47	524
$C_6H_{10}O$ (cyclohexanone)	ECH	18	71	846
$C_6H_5COCH_3$	PB	12	58	463
CH_3CHO	B	6	31-52	178-208
$(CH_3)_2CHCHO$	s-MP	12	72	615
C_3H_7CHO	H	6	53	206
$C_4H_9CH(C_2H_5)CHO$	EO	7.3	75	650

[a]Standard reaction charge: 500 ml liquid ammonia, 1.5 mol (90%) KOH, and 24 mol acetylene.
[b]MB, 3-methyl-1-butyn-3-ol; MP, 3-methyl-1-pentyn-3-ol; DMH, 3,5-dimethyl-1-hexyn-3-ol; ECH, 1-ethynylcyclohexanol; PB, 3-phenyl-1-butyn-3-ol; B, 1-butyn-3-ol; H, 1-hexyn-3-ol; s-MP, secondary 4-methyl-1-pentyn-3-ol; EO, 4-ethyl-1-octyn-3-ol.

acetone to methyl butynol (MB). A typical mole ratio used was acetone (1.0):acetylene(2.5):ammonia(1.5):potassium hydroxide (0.17). The conversions to MB based on acetone and potassium hydroxide were 87% and 600%, respectively, with 5% of the acetone recoverable. However, due to inherent hazards of handling undiluted liquid acetylene commercially, it is doubtful if such a reaction system would have a great enough advantage over the ammonia system to justify its risk.

Liquid Ammonia-Alkali Metal Acetylides Route Almost concurrent with the development of the alkali metal hydroxide-liquid ammonia route was the use of catalytic amounts of alkali metals in liquid ammonia (ENI, Balducci) to produce alkynols [137,140,141]. This method is being used at Ravenna, Italy (SNAM process) to catalytically produce methyl butynol (MB) from acetone and acetylene, which, after semihydrogenation to methyl butenol, is dehydrated to isoprene (see Sec. 1.6.8). A commercial plant rated at about 33 million lb/yr polyisoprene has been reported to be in operation at Ravenna based on catalytic MB.

Struzenegger [142] (Hoffmann-La Roche) has disclosed a continuous process for producing alkynols using technology closely related to the sodium-liquid ammonia route (ENI-SNAM). The novel features of this invention relate primarily to the continuous separation of ammonia from the alkynol-sodio alkynol mixture through the use of a wiped-film evaporator under pressure. It is probable that this method is, or was, used at the Roche vitamin A-vitamin E facility at Nutley, New Jersey for the manufacture of key intermediates such as MB, dehydrolinalool, and dehydronerolidol.

A comparison of patent and literature data [129,137,142] shows that the alkali hydroxide route has a two- to five-fold productivity advantage over the alkali metal process, in addition to the safety and cost advantages of potassium hydroxide versus sodium.

Mechanism of Catalytic Ethynylation Tedeschi proposed [129,132] the following mechanism for catalytic ethynylation in polar media based on studies [9] with alkali metal hydroxide-acetylene and alkynol complexes:

$$1.\quad CH{\equiv}CH + NH_3 \rightleftharpoons H_3N \longrightarrow H\text{-}C{\equiv}CH \overset{NH_3}{\rightleftharpoons}$$

$$H_3N \longrightarrow H\text{-}C{\equiv}C\cdots H \longleftarrow NH_3$$

$$\downarrow\uparrow$$

$$H_3N \longrightarrow \underset{(a)}{H\text{-}C{\equiv}C^-} + NH_4^+$$

2. $KOH + (a) \rightleftharpoons \left[H_3N \longrightarrow \underset{(b)}{H\text{-}C{\equiv}C^-} \right] KOH$

3. $(CH_3)_2\,C{=}O \; + \; (b) \rightleftharpoons$

$$\left[(CH_3)_2\text{-}\underset{O^-}{\underset{|}{C}}\text{-}C{\equiv}CH \longleftarrow NH_3 \right] KOH$$

(c)

4. $(c) + (a) \rightleftharpoons (b) + (CH_3)_2\text{-}\underset{O^-}{\underset{|}{C}}\text{-}C{\equiv}CH \longleftarrow NH_3$

(d)

5. $(d) + NH_4^+ \rightleftharpoons NH_3 + (CH_3)_2\text{-}\underset{OH}{\underset{|}{C}}\text{-}C{\equiv}CH \longleftarrow NH_3$

Step 3 is essentially nonreversible at 25 to 30°C, and only above 60°C is the carbinol-KOH adduct gradually decomposed to acetylene, ketone, and hydroxide. Also, the NaOH or KOH-methyl butynol complexes were observed to undergo solid-phase transformation on standing into the alkyndiol-base complex via the following route [9,10] (see also Sec. 2.15.3):

$$2 \left| (CH_3)_2\text{-}\underset{OH}{\underset{|}{C}}\text{-}C{\equiv}CH \right| KOH \longrightarrow \left| (CH_3)_2\text{-}\underset{OH}{\underset{|}{C}}\text{-}C{\equiv}C\text{-}\underset{OH}{\underset{|}{C}}(CH_3)_2 \right| KOH$$

MB-KOH DH-KOH

$+ \; KOH(C_2H_2)$

The NaOH complex was observed to disproportionate more rapidly than the KOH adduct. However, it was shown that the base-diol adduct (DH-KOH) was as effective a catalyst for ethynylation as the alkynol adduct (MB-KOH), hence cancelling any possible loss of catalyst activity [9,10,129,132,138].

It is likely that catalytic ethynylation with alkali metals (via the acetylides) probably proceeds through a carbanion mechanism similar to the alkali hydroxide system, except that an alkoxide or acetylide derivative of MB are the probable intermediate catalyst species:

$$2NaC_2H + (CH_3)_2C{=}O \longrightarrow \underset{\text{(a)}}{(CH_3)_2\underset{\displaystyle |}{\overset{}{C}}(ONa){\equiv}CH} + \underset{\text{(b)}}{(CH_3)_2C(OH)\text{-}C{\equiv}CNa}$$

The reaction of either Na or K dispersion (toluene) with MB yields, on isolation, unstable alkali metal derivatives [143] which, on reaction with methyl iodide, yield approximately equal amounts of the methylated derivatives of (a) and (b), supporting the existence of such species in alkali metal catalyzed ethynylations.

Miscellaneous Catalytic Ethynylation Systems *Basic resins* The use of quaternary ammonium hydroxide resins (Amberlite or Amberlyst types) has been described by Whitfield (ICI) [144], and later by Frantz [145], for the ethynylation of polar ketones. The method is a potentially useful continuous process with reactive ketones such as acetone and methyl ethyl ketone but gives low conversions (2-10%) with higher ketones. Reported mole conversions of acetone or MEK to MB and MP average 35 to 50% of theory. Aldehydes fail to react. A BASF patent [146] claims a more active resin catalyst made by impregnating the resin with catalytic amounts of KOH and the desired alkynol in methanol solution. Aldehydes also react, but in lower conversions, to form secondary alkynols.

The resin system is best operated as a continuous liquid-phase process at 40 to 120°C and hydrostatic pressures up to 1000 psig.

Polar solvents Nedwick [147] has described a continuous liquid-phase system for carrying out reactions such as ethynylation and vinylation. By the utilization of acetylene in a solvent such as methanol (27 wt %) cyclohexanone could be converted to ethynyl cyclohexanol in 51 to 54% conversion. The method, however, gave poor results with acetone and other ketones. Although more desirable solvents (N-methylpyrrolidone, methylal, dioxane, etc.) were mentioned, they were not employed.

Shachat and Bagnell (Rohm and Haas) [148] reported the catalytic ethynylation of ketones with either alkali metal acetylides, alkoxides, or hydroxides in a wide variety of solvents. Reasonably high conversions, however, were obtained only when highly polar solvents such as DMSO or NMP were used.

Blumenthal [149] also has described the catalytic formation of alkynols using sodium or potassium hydroxides in polar solvents such as dimethyl sulfoxide, acetonitrile, and ethylene diamine, carrying out the ethynylations at atmospheric pressure. A similar technique has recently been patented [150] by Rhone-Poulenc (France) for the preparation of the perfumery and vitamin A-E intermediate dehydrolinalool using aprotic solvents. This new atmospheric pressure process using NaOH is, or was, employed by Rhodia (owned by Rhone-Poulenc) in their new terpene chemicals plant (Freeport, Texas) [151] where methyl hepteneone (produced from isoprene and acetone) is reacted with acetylene to produce dehydrolinolool.

A vapor-phase method (Nogaideli) has been reported [152] for the ethynylation of acetone using a spheroidal catalyst comprising sodium hydroxide and gumbrin clay. Optimum conversions were obtained at 120 to 125°C (13-23%). This process is claimed to produce only methyl butynol. No details have been disclosed regarding catalyst life or the versatility of the method for other alkynols. Based on extensive published data concerning ethynylation, it is doubtful if a vapor-phase route can compete with known liquid-phase processes for producing a variety of alkynols and alkyndiols.

Technology Details: ANIC Catalytic Ethynylation Process The ANIC (SNAM-Progetti) continuous ethynylation process [137,140] for the production of methyl butynol is carried out in the activating media of liquid ammonia (see Sec. 1.6.8). This technology is also capable of producing a variety of secondary and tertiary acetylenic alcohols (see Secs. 2.15 and 2.16) by the ethynylation of aldehydes and ketones using catalytic amounts of potassium hydroxide. ANIC is now a recognized leader in the production of 3-methyl-1-butyn-3-ol(MB) and 3-methyl-1-buten-3-ol(MBe), the semihydrogenated derivative of MB. The dehydration of MBe is also used to produce high-purity isoprene in high yield (98%) at the ANIC production facility at Ravenna, Italy.

Figure 2-3 is a flow diagram of the ANIC process. The following feeds are premixed before entering the reactor: acetone, acetylene, ammonia-acetylene recycle, and 50% aqueous KOH. The reactor is stated to be a tubular, cooled reactor of the plug-flow type. It is also claimed that the use of aqueous KOH results in a homogeneous reaction mixture and a simple, efficient caustic feed system.

The original ANIC (ENI) process [137,140,141] used sodium dissolved in liquid ammonia as catalyst. The sodium, in the presence of excess acetylene, is rapidly converted to sodium acetylide (NaC_2H), which functions as the initial catalyst species (see discussion of the mechanism of catalytic ethynylation in this section). The sodium-based route is no longer employed since the use of KOH has proven to be cheaper, safer, and more efficient.

An excess of acetylene over acetone (acetylene-to-acetone mole ratio, 1.5-2.0) is maintained in the reactor to avoid the formation of acetone (condensation) by-products. Potassium hydroxide is used in catalytic amounts (0.017-0.071 mol) per mol of acetone, using 50% aqueous KOH as catalyst feed solution. Liquid ammonia functions as a polar solvent, and broad ranges are claimed in the SNAM patent [137,140]. Depending upon the feed composition used, ammonia composes approximately 30 to 50% (by weight) of the reaction mixture. Except for patent examples cited and broad ranges claimed, there are no further details published regarding optimum operating conditions. The reaction is carried out at 10 to 41°C and pressures of 280 to 350 psi in the liquid phase.

The reactor effluent is treated with either ammonium sulfate or chloride to neutralize potassium hydroxide catalyst and to prevent the reversal of methyl butynol to acetone, acetylene, and acetone condensation products. Ammonia and acetylene are recovered in a flash tank, where they are then separated under vacuum and recycled. In the fractionation columns shown (Fig. 2-3), crude MB is separated from acetone (overheads), and the MB bottoms are then purified in the second still to methyl butynol suitable for partial hydrogenation to methyl butenol. Since this grade of MB contains water (about 5-10%), further purification via extractive or azeotropic distillation is required for high-purity MB containing less than 0.1% water. The complete ANIC process for MB, MBe, and isoprene is discussed in Sec. 1.6.8.

Continuous Reaction System for Acetylene under Pressure in the Liquid Phase Nedwick [147] has described a safe continuous liquid-phase reaction system for bench- and pilot-scale acetylene reactions under pressure in the absence of an acetylene gas phase. The principles utilized in this earlier (1962) work are now used in present-day ethynylation processes such as the ANIC methyl butynol process. The absence of an acetylene gas phase helps prevent the possibility of an acetylene deflagration or detonation. A flow diagram of the reaction system is shown in Fig. 2-4.

The important components of this system are the feed reservoir, acetylene absorbing vessel, catalyst feed solution, coil reactor, post cooling coils, back pressure (let-down) regulator valves, product isolation vessel, and acetylene exit meter. The absorber is a 1-liter stirred and jacketed autoclave with an internal cooling coil. The hollow-shaft stirrer used is specially designed for gas dispersion, and the head of the autoclave is fitted with a capacitance liquid-level probe, the output of which controls the feed pump via a level controller. Acetylene metering is controlled by a differential pressure (d.p.) cell, and the system is calibrated at different pressures to determine the moles of acetylene introduced via a pneumatic square root integrator. Gaseous acetylene under pressure (300-500 psig) is added only to the absorber vessel, and

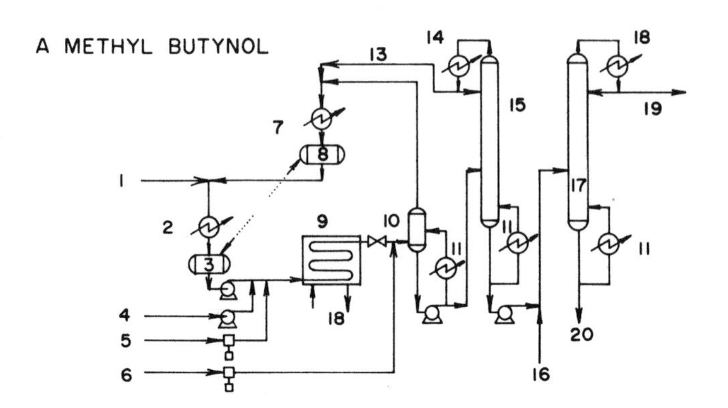
A METHYL BUTYNOL
1
2
3
4
5
6
7
8
9
10
11
13
14
15
16
17
18
19
20

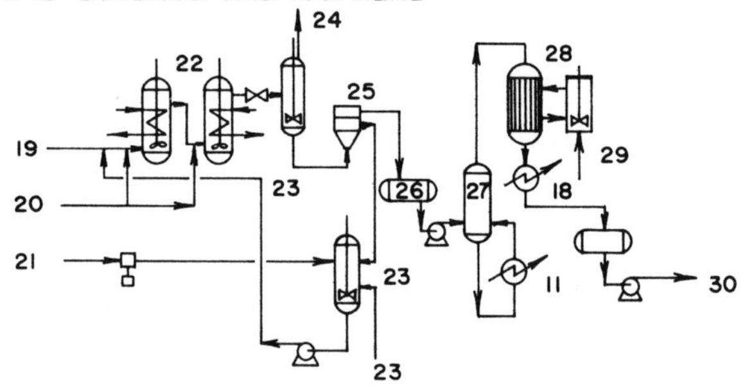
B METHYL BUTENOL AND ISOPRENE
18
19
20
21
22
23
24
25
26
27
28
29
11
30

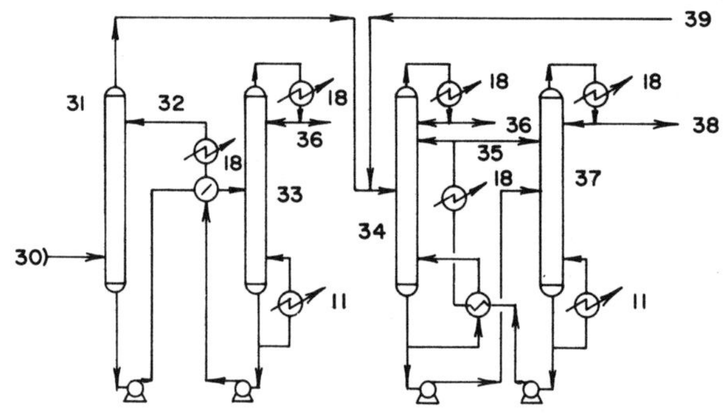
C ISOPRENE PURIFICATION
11
18
30
31
32
33
34
35
36
37
38
39

low temperatures (-20 to -40°C) are used to maintain safety (prevent detonation) and maximize acetylene solubility.

The process is started by pumping solvent, reactants, and acetylene to the absorber after the back-pressure (let-down) valve is set at 1500-2000 psig and the reactor bath is set at the desired temperature. The catalyst solution is pumped to a point at the top of the reactor, where it mixes with incoming feed from the absorber. The high pressure used insures a complete liquid phase even at temperatures of 130 to 250°C. An additional feed system for one of the reactants (ketone, for example) can be readily installed and used in a manner similar to the catalyst solution. The reactor effluent is cooled, passes through the let-down valve, and is brought to atmospheric pressure. Unreacted acetylene is determined via a wet-test flow meter.

The major advantages claimed for the continuous liquid-phase process are:

1. A dangerous gas phase of acetylene at higher temperatures and pressures leading to deflagration or detonation is avoided, thereby increasing process safety.
2. The use of polar solvents (DMF, N-Methylpyrrolidone, dioxane, etc.) facilitates the dissolving and stabilizing of acetylene and results in higher product yields.
3. Reaction rates and productivity are high, due to the excellent heat transfer characteristics of the reactor and the use of higher reaction temperatures (130-250°C). Also, the rate of reaction is not limited by the rate of acetylene transport across a gas-liquid boundary, further increasing reaction rate and conversion to product.
4. The reactor configuration is simple, small and designed for use at high hydrostatic pressure. Acetylene concentration under continuous conditions in the reactor is small compared to a batch system, thereby resulting in safer operation.
5. The use of a high hydrostatic pressure in the reactor (1500-2500 psig) eliminates the dangerous formation of an acetylene gas phase at higher temperatures.

Figure 2-3 ANIC (SNAM-Progetti) catalytic liquid ammonia process methyl butynol (Figure 1-16). (Reprinted by special permission from *Chemical Engineering*, October 1, 1973. Copyright 1973 by McGraw-Hill, Inc. New York.)

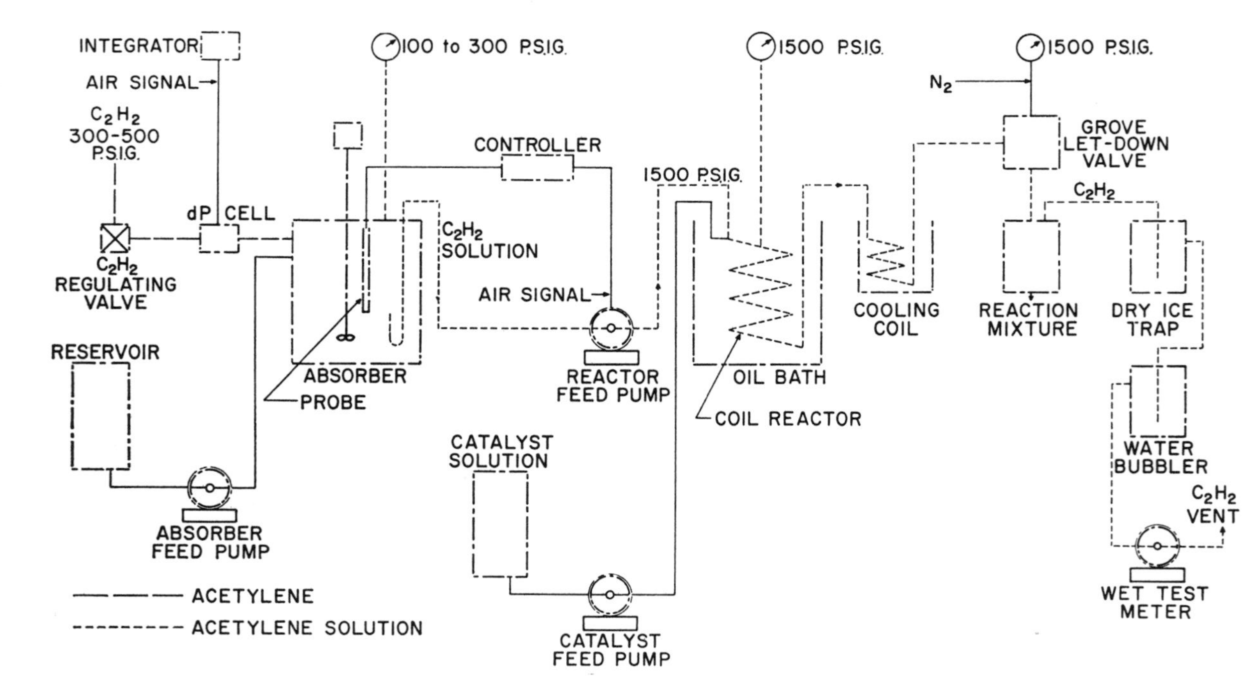

Figure 2-4 Continuous liquid-phase acetylene reactor. (Reprinted with permission from John J. Nedwick, "Liquid phase process for *acetylene reactions, I and EC Process Design and Development*, Vol. 1, April 1962, pp. 137-141. Copyright 1962 American Chemical Society.

6. Residence times in the reactor are very short (3-4 min), and consequently, the formation of polymers (polyacetylene and polyaldols) and other by-products is greatly reduced.

The versatility, safety, and high productivity of the reaction system has been demonstrated by the following reactions, which are detailed in Nedwick's publication:

a. Vinylation of alcohols, mercaptans, and pyrrolidone at 180 to 220°C and 1500-2500 psig using KOH as catalyst (average conversion 85-95%).
b. Ethynylation of cyclohexanone to 1-ethynylcyclohexanol at 135°C and 1500 psig in methanol using KOH as catalyst (conversion 51%).
c. Condensation of acetylene with diethyl carbonate using sodium ethoxide or Triton B ethoxide as catalysts at 195°C and 1500 to 2500 psig to yield ethyl-3-ethoxyacrylate and ethyl-3,3'-diethoxypropionate (total conversion 50-55%).

The contact times in the reactor average 3 to 4 min, indicating the rapid rate of reaction and the high productivity of the system.

Hoffmann-La Roche Catalytic Alkynol Process The Roche ethynylation process [142] is a variant of the original ENI catalytic liquid ammonia process [137,140,141] using catalytic amounts of sodium as catalyst under pressure. The novel use of a thin (wiped)-film evaporator for rapid flashing of ammonia and the concentration of the alkynol prior to final distillation has added further utility to the process. The process can be employed for the manufacture of such important alkynols as methyl butynol (MB) and dehydrolinalool (DHL), as well as higher polyene alcohols, which are important intermediates for the production of vitamins A, E, and K, as described in the Roche patent [142].

Figure 2-5 is a schematic of the Roche process (A) and the important wiped-film evaporator (B) as detailed in the Roche patent [142]. The important components of this continuous liquid-phase process are the reactor (plug flow), pressure control valves (PCV), preheater, wiped-film evaporator, condenser and splitter, and recycle line. The reaction is carried out by pumping separate feeds of (a) sodium solution in liquid ammonia, (b) acetylene, and (c) ketone, to the reactor (not shown in the schematic flow diagram). Acetylene can be conveniently dissolved

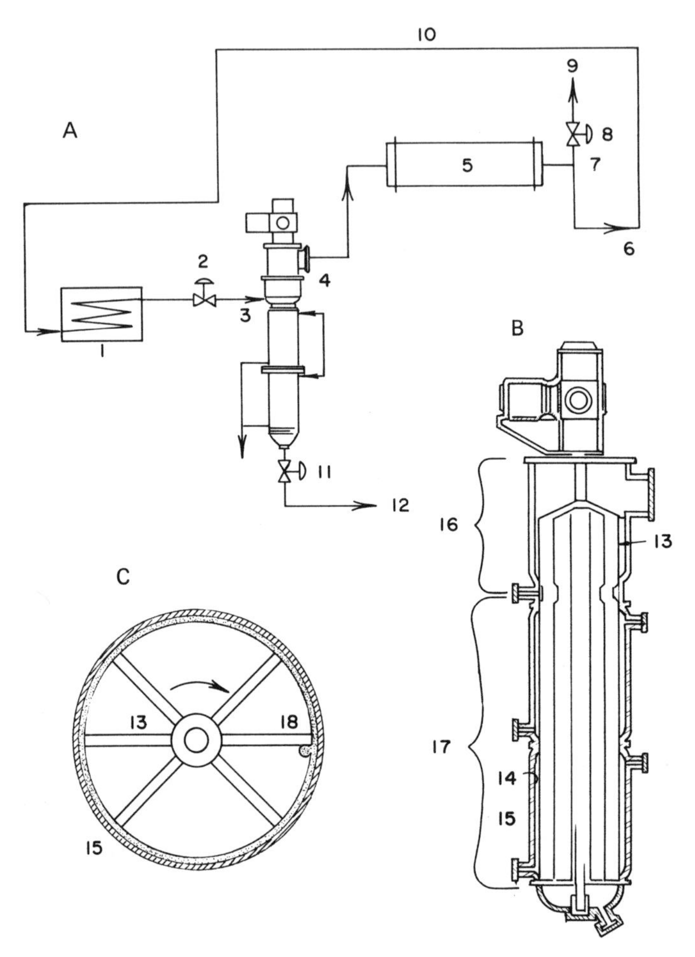

1. Ethynylation pressure reactor
2. Pressure control valve (PCV)
3. Preheater
4. Wiped-film evaporator
5. Condenser (liquid NH_3 + C_2H_2)
6. Ammonia, etc., condensate
7. Splitter
8. PCV to vent
9. Vent
10. Recycle line
11. PCV from evaporator (4)
12. Product effluent
13. Film wiper (evaporator) blade
14. Interior condensing (film) surface
15. Exterior shell of (4)
16. Separating section
17. Thermal section
18. Liquid and fillet

in the liquid ammonia containing the sodium, or fed to the reactor separately.

Sodium and acetylene in liquid ammonia react readily to form sodium acetylide and hydrogen, with a minor amount of the acetylene reduced to ethylene and ethane. These inerts are removed by the vent valve (11). The sodium acetylide, on reaction with ketone, forms the sodio derivative of the alkynol, which is the active catalyst species for the ethynylation.

$$Na + C_2H_2 \xrightarrow{NH_3} NaC{\equiv}CH + \tfrac{1}{2}H_2$$

$$R_1R_2CO + NaC{\equiv}CH \longrightarrow R_1R_2C(ONa){-}C{\equiv}CH$$

It is also likely that the disodio derivative of the alkynol formed by the reaction with ethynyl hydrogen is also present as a catalyst specie. The key feature of the La Roche process is the use of the wiped-film evaporator at elevated temperatures (70-120°C), using short contact times and condensing the liquid ammonia under pressure at about 200 psig. The temperature gradient from the top to the bottom of the evaporator varies from 20 (top) to 120°C (bottom). The speed and efficiency of acetylene-ammonia evaporation (flashing) and condensation in the wiped-film evaporator (short residence time) prevents significant base (sodio derivatives)-catalyzed cleavage of the alkynol product back to ketone and acetylene and subsequent ketone condensation products.

Details of product isolation are not given in the Roche patent [142], but it is likely that the crude product emerging as bottoms (14) from the evaporator is cooled; neutralized with either acid (acetic, phosphoric, etc.), ammonium salts, or carbon dioxide; filtered; and then fractionally distilled. The following reaction charge and conditions summarize the catalytic formation of 3-methyl-1-butyn-3-ol by the reaction of acetone and acetylene using sodium as catalyst in liquid ammonia: (a) 1.04 kg acetylene (40.0 m), (b) 0.023 kg sodium (1.0 m), equivalent to 0.048 kg NaC_2H, (c) 3.97 kg liquid ammonia (233.5 m), (d) 1.66 kg acetone (28.6 m).

Figure 2-5 Hoffmann-La Roche continuous ethynylation process with wiped-film evaporator: (A) flow diagram, (B) vertical section of (4), (C) horizontal section of (4).

Reaction conditions: temperature, 5°C; pressure, 700 psig
Pumping rates: (a) + (b) + (c), 450 g/hr; acetone, 144 g/hr
Reactor effluent: ammonia, 3.97 kg; acetylene, 0.33 kg; methyl butynol, 2.16 kg; acetone, 0.08 kg; sundries, 0.15 kg
Condensation (NH_3 + C_2H_2): temperature, 35°C; pressure, 215 psig
Bottom evaporator temperature: 118°C
Yield (selectivity) to methyl butynol: 90%

Liquid Acetylene: A Batch Reaction System Tedeschi and coworkers [133] have described a bench-scale batch system for the safe liquification of acetylene and for carrying out a variety of synthetic reactions using liquid acetylene as reactant and solvent at or below its critical temperature (35°C). Two reactors were employed behind a special barricade equipped with explosion-proof visual ports. One reactor (rocker type) had two glass walls to permit visual inspection of liquid acetylene reactions under pressure, while the second reactor was a stirred, double-walled autoclave (125-ml capacity) in which heat-exchange fluid circulated between walls for rapid heating and cooling.

Both reactors could be operated at the same time, and the temperature varied from -60 to 60°C in less than 15 min. The reaction system was used safely for over 6 months on a daily basis in the formation of hydrogen-bonded [133] and transition metal complexes [139], alkali metal (Li, Na, K) acetylides, sodium propiolate [153], propargyl amines, diaminobutynes, and acetylenic alcohols (catalytic ethynylation [139]. The use of liquid acetylene enabled these reactions to proceed rapidly in good-to-excellent yields at lower temperatures with fewer by-products.

Presented below are flow diagrams of the reaction system (Fig. 2-6) and the various components, such as the accumulator oil system for compressing acetylene, sight-glass reactor and heating-cooling tank, heating-cooling system, and panelboard layouts for operation of the various components (Fig. 2-7). Although this work gave promising and interesting results, it is not a recommendation to use liquid acetylene in synthesis operations either commercially or at the bench-scale level without adequate safeguards to prevent and contain explosions. The hazards of handling acetylene under pressure are discussed in Secs. 1.4.7 to 1.4.10. However, this reaction system, due to its versatility, can also be used for solvent acetylene-based reactions, as well as other synthetic operations such as hydrogenation, ethoxylation, or reactions requiring the use of liquified gases.

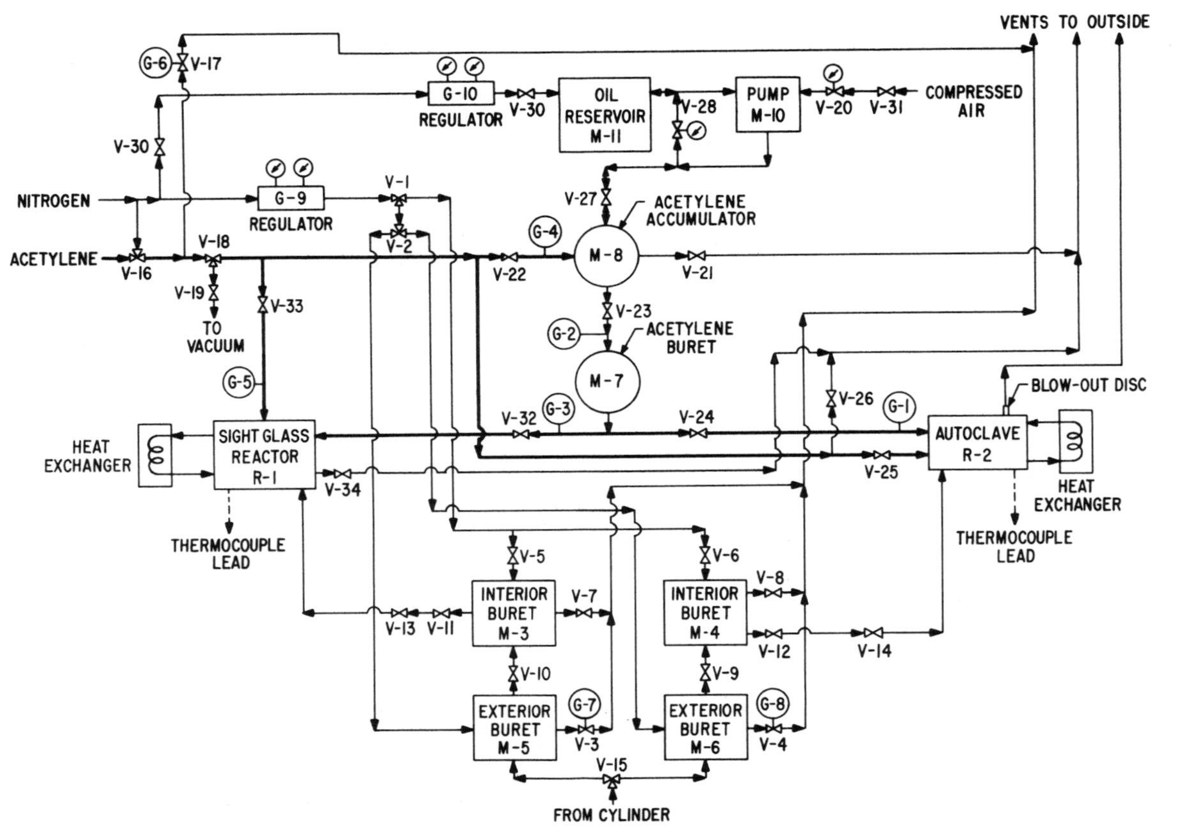

Figure 2-6 Batch reaction system for liquid acetylene reactions. (Reprinted with permission from Tedeschi, et al., Liquid acetylene, *I and EC Process Design and Development*, Vol. 7, April 1968, pp. 303-307. Copyright 1968 American Chemical Society.)

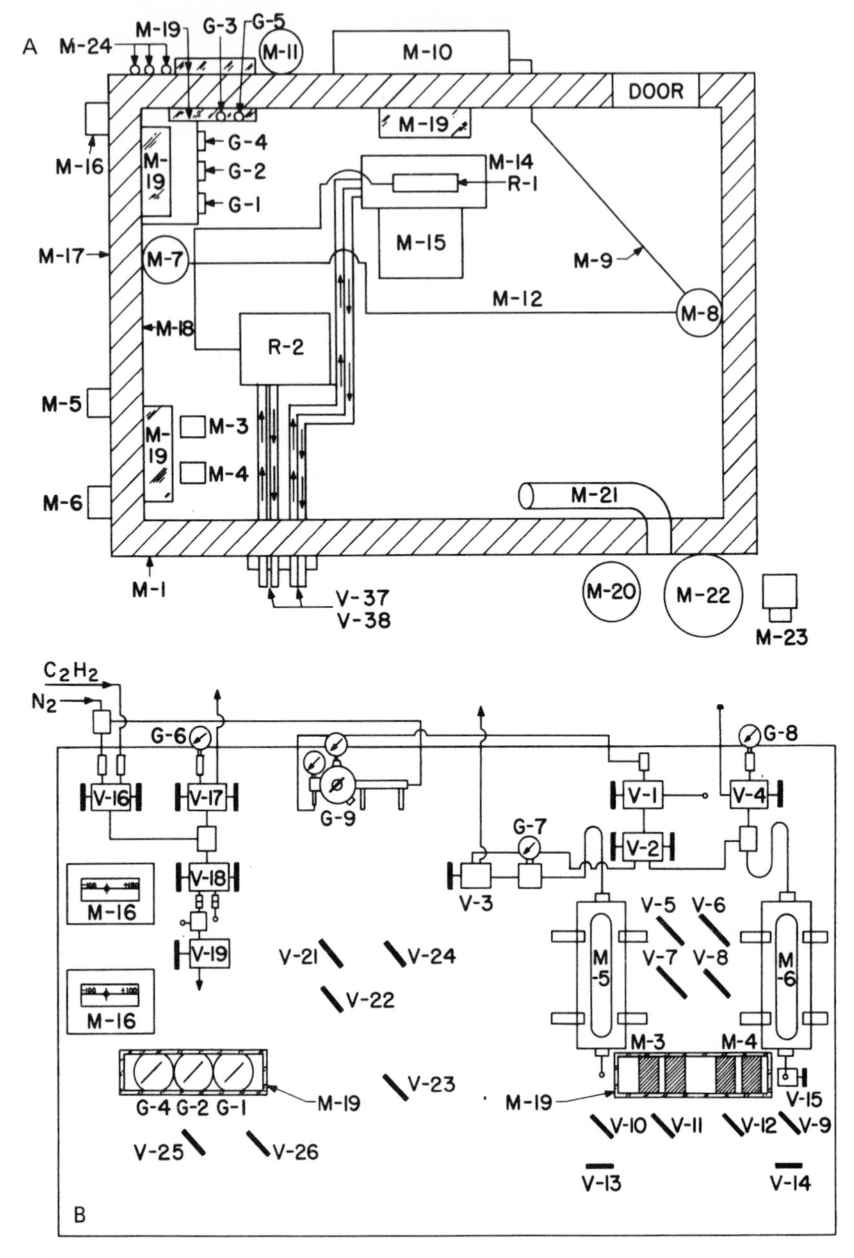

Figure 2-7 Batch reaction system for liquid acetylene reactions. (Reprinted with permission from Tedeschi, et al., Liquid acetylene, *I and EC Process Design and Development*, Vol. 7, April 1968, pp. 303-307. Copyright 1968 American Chemical Society.) (A) Top view of cubicle; (B) Outer front (main) panel board.

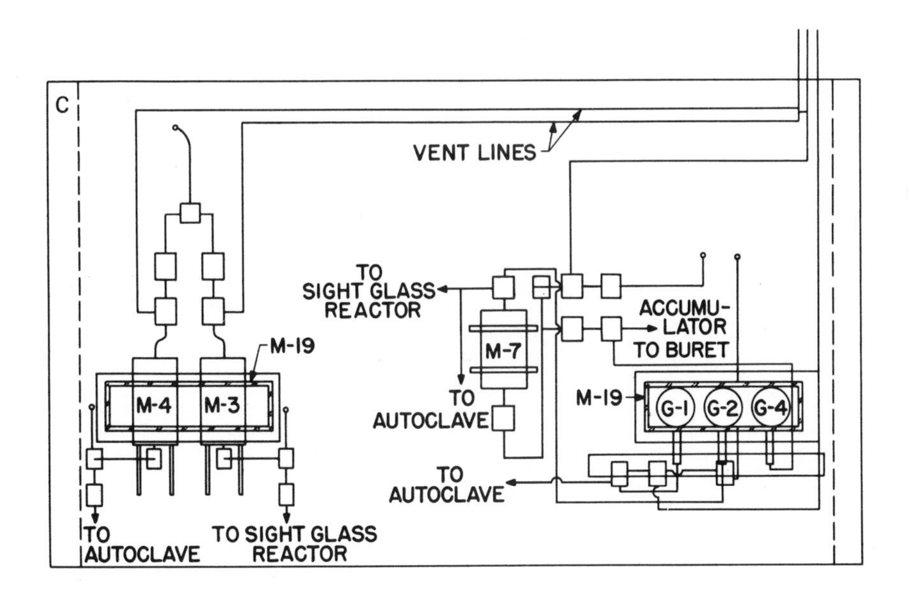

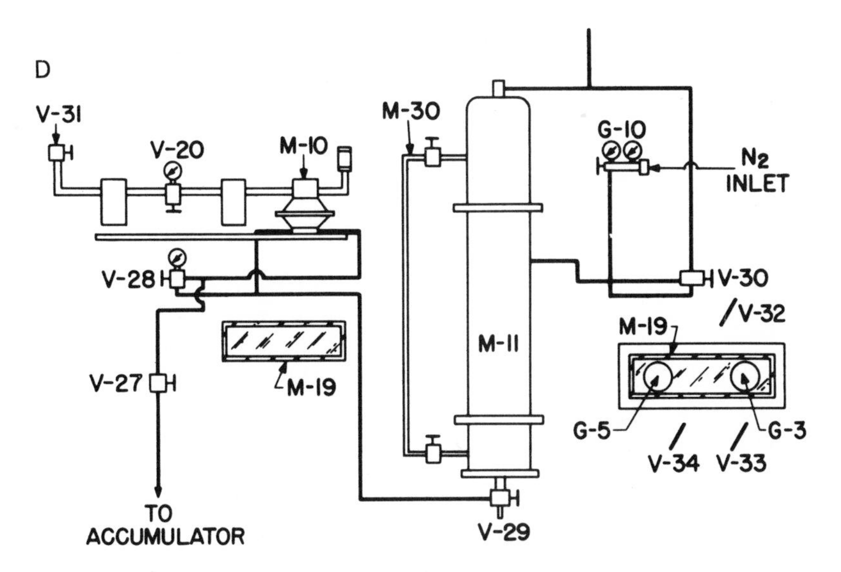

Figure 2-7 (C) Inner front panel board; (D) Accumulator system.

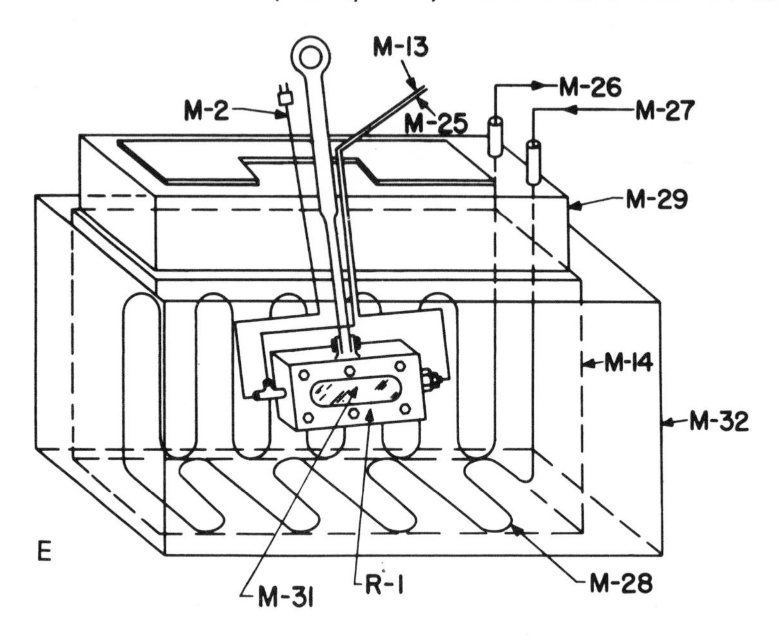

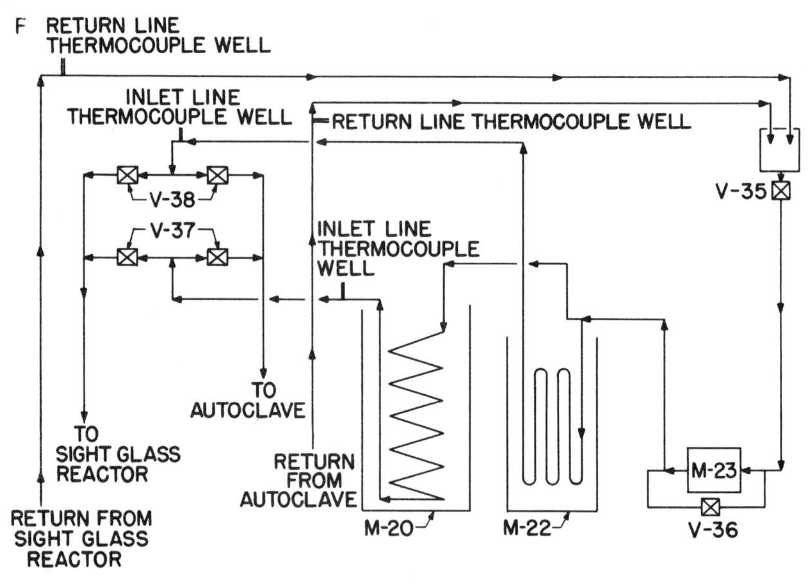

Figure 2-7 (E) Sight glass reactor and heating-cooling tank; (F) Heating-cooling system.

2.16 COMMERCIAL USES FOR SECONDARY AND TERTIARY ACETYLENIC ALCOHOLS AND DIOLS

During the early 1960s the Air Reduction Company developed a line of specialty acetylenics based on the reaction of aldehydes and ketones with acetylene. These products were not competitive with Reppe technology (propargyl alcohol and butyndiol) and served to introduce a versatile line of alkynols and alkyndiols to the U.S. marketplace.

On January 28, 1971 the assets and business of the Airco Chemical Division of the Air Reduction Company (AIRCO) were purchased by Air Products and Chemicals, Inc. (APCI), and their acetylenics chemical line was subsequently set up as a separate profit center, the Acetylenic Chemicals Division, in the Houdry Division of APCI. In recent years APCI acetylenics have been combined with other Houdry specialty products into a new corporate structure, known as Performance Chemicals.

2.16.1 Building Blocks

The largest-volume alkynol today is methyl butynol (MB), due primarily to large-volume uses in the manufacture of isoprene (see Sec. 1.6.8) and vitamins A and E [154,155]. The SNAM-Progetti production facility at Ravenna, Italy has been rated [154] initially at 33 million lb/yr isoprene, which, in turn, would logically indicate MB production substantially greater than the isoprene value. This plant has been expanded since it was built, but present-day capacity has not been announced. It is expected that this cheap source of methyl butynol will have a noticeable effect on the marketplace in the years ahead and could stimulate new uses for this versatile alkynol.

Total U.S. production of methyl butynol for the manufacture of vitamins A and E has been estimated during 1971 [156] at 3 to 5 million lb annually, with worldwide consumption at possibly twice this volume. MB produced in bulk is a commercially attractive chemical for use in metal treatment, stabilization of chlorinated hydrocarbons, as a synthesis building block, and in specialty solvent applications (Table 2-8).

Other tertiary alkynols such as 3-methyl-1-pentyn-3-ol (MP) [155] and 1-ethynylcyclohexanol (ECH) [157] are believed to be low-volume specialties at present. MP was one of the first acetylenic alcohols produced in the United States (early 1950s) and was marketed as a tranquilizer under the trademark Dormison. ECH is a potentially cheap alkynol based on cyclohexanone that has potential as a building block, corrosion inhibitor, and stabilizer

Table 2-8 Acetylenic Alcohols and Derivatives

Products	Structural formula
Methyl butynol (MB)	$CH_3\text{-}C(CH_3)(OH)\text{-}C{\equiv}C\text{-}H$
Methyl pentynol (MP)	$CH_3\text{-}CH_2\text{-}C(CH_3)(OH)\text{-}C{\equiv}C\text{-}H$
Ethynyl cyclohexanol (ECH)	(S) ring bearing $C{\equiv}C\text{-}H$ and OH
Hexynol (H)	$CH_3\text{-}CH_2\text{-}CH_2\text{-}CH(OH)\text{-}C{\equiv}C\text{-}H$
Ethyl octynol (EO)	$CH_3\text{-}CH_2\text{-}CH_2\text{-}CH_2\text{-}CH(CH_2\text{-}CH_3)\text{-}CH(OH)\text{-}C{\equiv}C\text{-}H$
OW-1	Proprietary Product

Boiling Point (°C, 760 mm)	Melting Point (°C)	Solubility in H_2O (wt, % at 20°C)	Applications
103.6	2.6	miscible	Stabilizers in chlorinated solvents; viscosity reducers and stabilizers; electroplating brighteners; intermediate in synthesis of hypnotics and isoprenoid chemicals such as vitamin A, ionone, and perfume alcohols; solvents for alcohol-soluble nylon and polyamide resins
121.4	-30.6	9.9	
180	30-31	1.2	Corrosion inhibitor for mineral acids; stabilizer in chlorinated organics; synthesis of hypnotics, other pharmaceuticals, and perfumery materials
142	-80	3.8	Corrosion inhibitor for mineral acids; high temperature oil-well acidizing inhibitor
197.2	-45	<0.1	Acid inhibitor for mild steel, acid pickling and cleaning baths, electrolytic cleaning baths, electrolytic refining of metals, acidizing of oil wells
		<0.1	High- and intermediate-temperature corrosion inhibitor for mineral acids in oil-well acidizing and metal cleaning applications

Table 2-8 (Continued)

Products	Structural formula
Dimethyl hexynediol (DH)	$CH_3-C(CH_3)(OH)-C{\equiv}C-C(CH_3)(OH)-CH_3$
Dimethyl hexanediol (DHS)	$CH_3-C(CH_3)(OH)-CH_2-CH_2-C(CH_3)(OH)-CH_3$
Dimethyl octanol	$CH_3-CH_2-C(CH_3)(OH)-CH_2-CH_2-CH(CH_3)-CH_2-CH_3$

for chlorinated solvents. Applications for secondary and tertiary acetylenic alcohols, tertiary acetylenic diols, and derivatives are summarized in Table 2-8.

2.16.2 Corrosion Inhibitors, Intermediates, and Fragrance Compounds

The most important use for secondary acetylenic alcohols (Table 2-8) that has evolved in recent years is corrosion inhibition at ferrous surfaces in the presence of mineral acids. Commercial representatives of this class are 1-hexyn-3-ol (H) [158] 4-ethyl-1-octyn-3-ol (EO) [158], OW-1 (proprietary) [159] and propargyl alcohol (1-propyn-3-ol) [21]. These products are currently being used in oil-well acidizing operations, where hydrochloric acids containing formulated acetylenic alcohols are pumped "down-hole" into steel tubing, to depths often in excess of 20,000 ft, to dissolve

Boiling Point (°C, 760 mm)	Melting Point (°C)	Solubility in H_2O (wt, % at 20°C)	Applications
205-6	94-95	27.0	Component in wire drawing lubricant formulations, coupling agent in resin coatings, organic synthesis
214-15	87-88.5	14.3	Synthesis of peroxide catalysts and cross-linkers, cyclic musk compounds, and allethrin
202	-67.5	0.045	Intermediate in the synthesis of pharmaceuticals; linalool-type odor, used in floral odors and perfumes and soaps

and loosen lime deposits and free trapped oil. This oil well stimulation (primary oil recovery) is expected to grow in deep-hole applications as the use of arsenic (arsenites) is gradually phased out due to increased state and federal legislation against pollution of ground water and natural resources.

The tertiary acetylenic diol dimethyl hexyndiol (DH) [160] is a potentially cheap building block (based on acetone and acetylene) that has not been widely utilized at present. It has potential in metal treatment applications, besides utility as a versatile intermediate (Table 2-8). Its saturated analog dimethyl hexanediol (DHS) [161] is used industrially in the manufacture of polycyclic musks for the perfume industry and for specialty high-temperature diperoxides for the cross-linking of polymers (polyethylene, vinyls) in the manufacture of plastic pipe. The gem-dimethyl grouping in DH and DHS is believed to confer unique properties for both applications. Olefinic-1,4-diols, readily available via semihydrogenation of the corresponding acetylenic diols, are interesting products, containing essentially a cis(98%) configuration that has not yet been utilized commercially.

Dimethyl octanol (Table 2-8) is a perfumery compound [162] derived from acetylene technology. In general, an acetylenic alcohol possessing a chain length of 6 to 12 carbons, together with branched methyl or gem-dimethyl groups, yields interesting odor properties. The olefinic and saturated analogs appear to have superior fragrance properties versus the acetylenic alcohols.

2.16.3 Alkyl Acetylenes (1-Alkynes)

The 1-alkynes of greatest commercial importance are methyl and ethyl acetylenes. These products (Table 2-9) at present are used almost exclusively as building blocks for the manufacture of pharmaceuticals and perfume-flavor products. Although relatively expensive as acetylenics, both methyl and ethyl acetylenes [163] are potentially cheap chemicals if recovered from petroleum refinery (cracking) streams in large volume. Specific areas of use are in oral contraceptives, anesthetics, and green odor (notes) for fragrance and flavor compounds. The formation of leaf alcohol (cis-3-hexene-1-ol) from ethyl acetylene is a well-known use for this alkyne [164]. Since leaf alcohol occurs widely in nature (grass, fruits), the synthetic hexenol has interesting potential as a flavor ingredient in a wide variety of applications.

2.16.4 Acetylenic Surfactants*

Tertiary acetylenic alcohols and glycols [160,167] in which the aliphatic chain is greater than six carbon atoms exhibit interesting wetting and defoaming properties (Table 2-10). The most important members of this unique surfactant class are Surfynol 104 (2,4,7,9-tetramethyl-5-decyne-4,7-diol) and Surfynol 82 (3,6-dimethyl-4-octyn-3,6-diol) [160,165]:

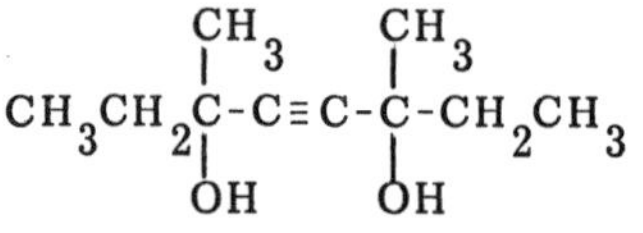

$$CH_3CH_2\overset{\displaystyle CH_3}{\underset{\displaystyle OH}{C}}\text{-}C\equiv C\text{-}\overset{\displaystyle CH_3}{\underset{\displaystyle OH}{C}}\text{-}CH_2CH_3$$

Surfynol 82

$$CH_3\overset{\displaystyle CH_3}{CH}CH_2\overset{\displaystyle CH_3}{\underset{\displaystyle OH}{C}}\text{-}C\equiv C\text{-}\overset{\displaystyle CH_3}{\underset{\displaystyle OH}{C}}\text{-}CH_2\overset{\displaystyle CH_3}{CH}CH_3$$

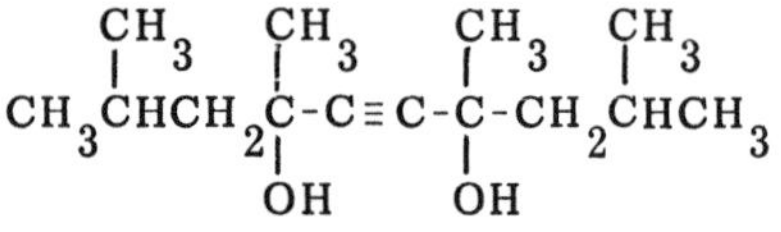

Surfynol 104

Surfynol 104 surfactant is also available commercially as ethoxylated surfactant derivatives known as the Surfynol 400 series [165, 166], which are produced in ethylene oxide to Surfynol mole ratios

*Surfynol: Air Products and Chemicals registered trademark for acetylenic surfactants.

Table 2-9 Alkyl Acetylenes (1-Alkynes)

Products	Structural Formula	Boiling Point (°C, 760 mm)	Melting Point (°C)	Specific Gravity and Vapor Pressure	Applications
Methyl Acetylene	$CH_3C{\equiv}CH$ (1-propyne)	-23.1	-101.5	0.660 (-13/4) 5.1 atm (20°C)	Special fuel applications, synthesis intermediate
Ethyl Acetylene	$CH_3CH_2C{\equiv}CH$ (1-butyne)	8.3	-122.5	0.669 (0/0 1.6 atm (20°C)	Intermediate for pharmaceutical and perfume-flavor compounds

varying from 3.5 to 30, and are commonly known as Surfynol surfactants (S-), 440, 465, and 485 (Table 2-10).

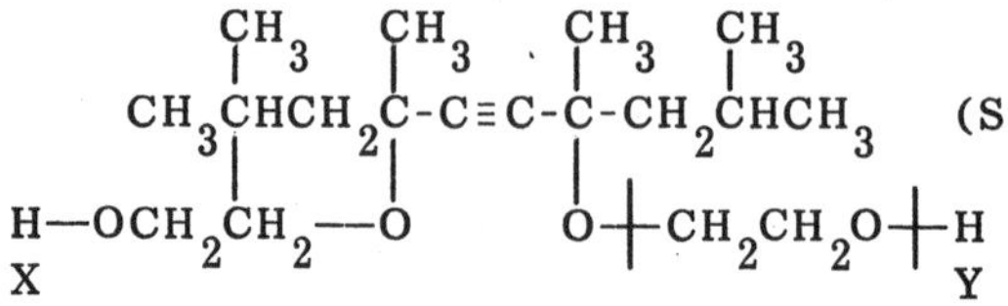

$$\begin{array}{c} \text{CH}_3\text{CH(CH}_3\text{)CH}_2\text{C(CH}_3\text{)(O(CH}_2\text{CH}_2\text{O)}_X\text{H)-C}\equiv\text{C-C(CH}_3\text{)(O(CH}_2\text{CH}_2\text{O)}_Y\text{H)CH}_2\text{CH(CH}_3\text{)CH}_3 \end{array}$$

(Surfynol 400 series)

The Surfynol 400 series, in the lower oxide ratios (S-440, S-450), are highly effective low-foaming wetting agents suitable for continuous wetting applications (water drop, defoamer, rinse aid, metal degreasing). The dual combination of a symmetrically situated triple bond flanked by adjacent hydroxyl or polyalkoxy groups makes the Surfynol products useful for metal treating and corrosion inhibition. Consequently, they have potential in such applications as boiler scale removal, acid pickling, metal degreasing, dip coating, electrodeposition of primer coats [167], metal fabrication, and wire drawing.

Surfynol 104 surfactant, although sparingly soluble (0.10%, 25°C), is one of the most rapid wetting agents available today. Its formulations 104-A and 104-E [168], Surfynol TG [169], Surfynol TG-E, and Surfynol PC [170] products are used as defoamers for paints, paper coatings, and textile finishes (104-A, 104-E, PC), as low-foam wetting agents for paints and pigmented aqueous systems (TG, 104-A, 104-E), and as agricultural wetting-defoaming formulation aids (104-A, TG-E). Surfynol surfactants 465 and 485 are used as synergistic surfactants [167] in the manufacture of latex paint emulsions, and confer such properties as freeze-thaw stability, improved pigment dispersion, and hiding [165]. They are also of value as adjuvants for use with pesticides.

Surfynol surfactant 82 (Table 2-10) is used as a cosmetic additive, as a defoamer in electroplating baths, and as an agricultural wetting aid [160,165]. Although not as powerful a wetter as Surfynol 104 product, its higher water solubility gives it greater versatility when used as a complexing and compatibility agent, which serves to modify the viscosity of formulations. Surfynol 61 (3,5-dimethyl-1-hexyn-3-ol) is used primarily as a volatile surfactant where only initial wetting is desired in a given process. Waterproofing of fabrics and fiber-glass cloth dyeing are typical application areas.

Surfynol SE surfactant and adjuvant is a formulation based on the Surfynol 104 product. It exhibits synergistic performance with other surfactants and additives in formulations. Like Surfynol 104 surfactant, it exhibits high-speed wetting with low foam, and in addition, forms stable aqueous emulsions up to 75% by volume. Its diverse applications are described in Table 2-10.

2.16.5 Surfynol 104 Surfactant and Oily Surfaces

Surfynol 104 surfactant, due to its unique structure (symmetrical triple bond with adjacent hydroxyl groups), gives rise to such interesting properties as metal surface affinity, high-speed wetting, defoaming capability, and complex formation [172,173]. Figure 2-8 illustrates the structure and surfactant properties of Surfynol 104 surfactant (S-104) and related acetylenic products. All Surfynol surfactants, and particularly the acetylenic alcohols, possess affinity for metal surfaces. This affinity is believed to involve the primary interaction of the pi field of the triple bond, reinforced by the adjacent polar hydroxy groups. Also shown in Fig. 2-8 are typical S-104-based formulations, which are useful in latex coatings.

When a Surfynol 104 product is formulated into typical water-based emulsions comprising, respectively, such systems as acrylic, polyester-alkyd, epoxy or vinyl acetate polymers, and the resulting formulation is applied to a greasy or oil-covered metallic (iron, stainless steel, aluminum) surface, significant performance improvements in the resultant baked or air-dried coating are noted. Coating improvements and performance characteristics are summarized in Table 2-11 with respect to Surfynol wetting capability. The more important benefits noted on oily or greasy metal surfaces are improved coverage, adhesion, leveling, water resistance, pigment wetting, and color development. Additional benefits related to wetting are surface tension reduction and spreadability, antistatic properties, and rinse-aid capability.

In Table 2-12 are shown other performance improvements related to defoaming and viscosity control problems encountered during formation of the finished coating. Defoaming capability, superior coverage on oily, greasy, and dirty surfaces, and sheeting action are considered the more important performance characteristics summarized in Tables 2-11 and 2-12.

2.16.6 Water-Based Coatings

The ability of Surfynol 104 surfactant in water-based coatings to increase the spreadability and overall performance of the coating on greasy or oily metal surfaces is believed to be related to the metal-Surfynol interaction proposed earlier. It is theorized that the strong attraction of the acetylenic diol molecule for the polar metal surface in effect displaces oil from the metal surface, enabling closer contact of the polymer emulsion with the metal

Table 2-10 Surface-Active Acetylenics

Trademark	Chemical Composition	% Active by Weight	Surface Tension (0.1% aq. soln., dyne/cm, 25°C)	Applications
Surfynol 104	Tertiary acetylenic glycol	100	31.6	Increased wetting and low foam with other surface active agents, defoamer in aqueous systems, rinse-aid surfactant
Surfynol 104A	Solution of 104 in 2-ethyl hexanol	50	33.0	Defoamer in emulsion systems for paints, paper coatings, and textile finishes; insecticide formulations; low-foam detergents
Surfynol 104E	Solution of 104 in ethylene glycol	50	36.2	
Surfynol TG	Proprietary Surfynol surfactant	83	27.6	Low-foaming wetting agent for latex paint and other pigmented aqueous systems
Surfynol 61	Dimethyl hexynol	100	32.4 (1.0% aq. soln.)	Volatile wetting agent for paper coatings, floor polishes and glass cleaning formulations; viscosity reduction
Surfynol 82	Dimethyl octynediol	100	55.3	Viscosity reduction in vinyl plastisols and aniline inks;

				cosmetic ingredient; low-foam wetting agent in developer compounds; defoamer in electroplating baths
Surfynol PC	Proprietary blend of an acetylenic glycol with other surface-active agents	100	—	Defoamer for paper coating latexes and systems where the foaming influence is a water-soluble polymer
Surfynol 440	Ethoxylated (3.5 mol) Surfynol 104	100	31.0	For water-based industrial finishes; lowers water-drop time of paperboard coatings; Dispersible in water, soluble in organic solvents
Surfynol 450	Ethoxylated (5 mol) Surfynol 104	100	31.5	For pigment dispersion in aqueous systems; dispersible in water, soluble in organic solvents
Surfynol 465	Ethoxylated (10 mol) Surfynol 104	100	33.2	For paint emulsion polymerization when blended with S-485, water -soluble, soluble in organic solvents, has high cloud point in presence of dissolved electrolytes or salts

Table 2-10 (Continued)

Trademark	Chemical Composition	Active by Weight	Surface Tension (0.1% aq. soln., dyne/cm, 25°C)	Applications
Surfynol 485	Ethoxylated (30 mol) Surfynol 104	100	40.1	For paint emulsion polymerization when blended with S-465, water-soluble, soluble in organic solvents, has high cloud point in presence of dissolved electrolytes or salts
Surfynol SE	Proprietary Surfynol 104-based blend		32.0	Water-based paint, inks, dye-processing, paper coatings, aqueous lubricants, metal treatment, pesticide formulation aid, adjuvant for pesticides

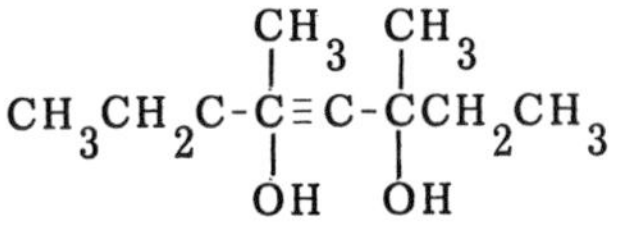

$CH_3CH_2C(CH_3)(OH)\text{-}C{\equiv}C\text{-}C(CH_3)(OH)CH_2CH_3$

Surfynol 82
(3,6-Dimethyl-4-Octyn-3,6-Diol)

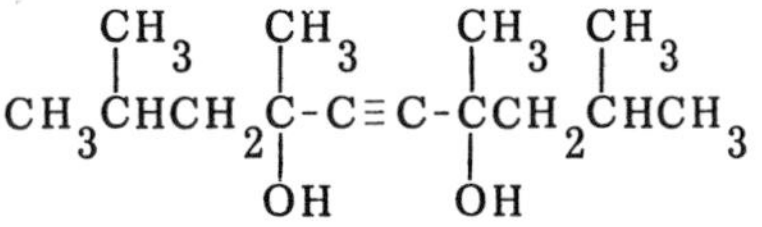

$CH_3CH(CH_3)CH_2C(CH_3)(OH)\text{-}C{\equiv}C\text{-}C(CH_3)(OH)CH_2CH(CH_3)CH_3$

Surfynol 104
(2,4,7,9-Tetramethyl-5-Decyn-4,7-Diol)

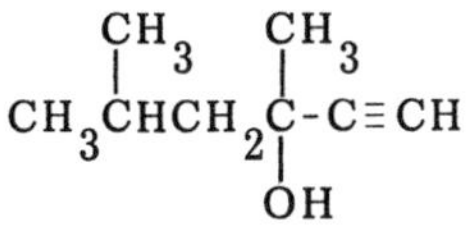

$CH_3CH(CH_3)CH_2C(CH_3)(OH)\text{-}C{\equiv}CH$

Surfynol 61
(3,5-Dimethyl-1-hexyn-3-Ol)

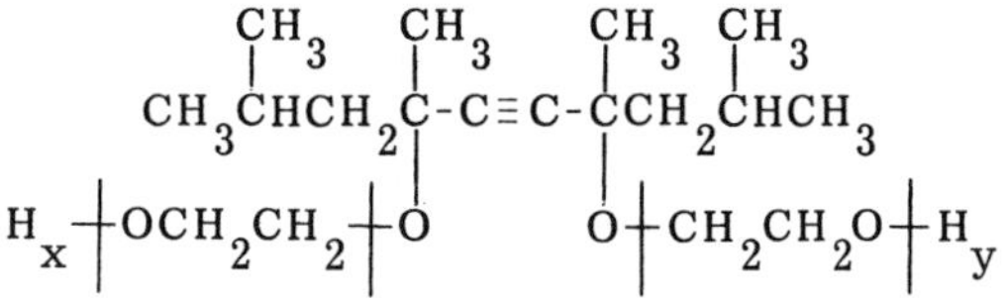

$CH_3CH(CH_3)CH_2C(CH_3)\text{-}C{\equiv}C\text{-}C(CH_3)CH_2CH(CH_3)CH_3$, with $H_x{+}OCH_2CH_2{+}O$ on the first substituted carbon and $O{+}CH_2CH_2O{+}H_y$ on the second

Surfynol 400 Series
(Surfynol 104 Ethoxylates)

Typical surfactant properties

Surfynol (S-)	HLB[a]	Wt. % Ethylene Oxide	Draves wetting time (secs)	Conc. wt. %	Foam
S-61	—	0	0	1.0	0
S-82	6	0	25	2.0	0
			0	3.0	0
S-104	4	0	25	0.07	0
			0	0.12	
S-440	6	40	25	0.07	0
			0	0.12	
S-465	13	65	25	0.20	Moderate
S-104-A[b]	4	0	31	0.08	
			0	0.24	0
S-TG-E[c]	10		8	0.1	Low
			0	0.2	

[a]HLB: Hydrophilic-lipopylic balance (estimated)
[b]S-104-A: 50% Surfynol 104 + 50% 2-Ethylhexanol
[c]S-TG-E: Proprietary Surfynol mix

Figure 2-8 Acetylenic Surfactants.

Table 2-11 Performance Characteristics (Wetting) of Coatings Containing Surfynol 104 Products

1. Superior coverage and adhesion on oily, greasy, or dirty surfaces
2. Improved leveling and reduction of cratering
3. Improved water resistance due to Surfynol volatilization during bake cycle
4. Improved recoatability vs. silicones
5. Enhanced pigment wetting and color development (S-TG)
6. Increased surface tension reduction and surface spreadability (high wetting speed)
7. Antistatic wetting properties
8. Improved sheeting action in rinse aids

surface. Surfynol 104 surfactant can also be regarded as a phase transfer agent, distributing itself between the latex, oil, and metal phases, thereby facilitating better metal coating contact. These effects are also manifest in the baking cycle, where coatings with significant reduced fish eyes, crawling, cratering, framing, edge pull, or flash rusting are obtained when Surfynol 104 surfactant is used.

In Table 2-13 is summarized typical performance improvements observed by the use of Surfynol 104 surfactant versus the same formulation without the Surfynol product. It can be seen that performance improvements similar to those desired in Tables 2-11 and 2-12 are realized with a variety of different coating systems.

Table 2-12 Other Performance Characteristics of Coatings Containing Surfynol 104 Products

Defoaming

1. Pigment grinding operations
2. Production, containerization and application of coating
3. Potentiation (increased wetting, defoaming) with other defoamers
4. Reduction of defoamer problems (fish-eyes, incompatibility)

Viscosity control

5. Reduction
6. Stabilization

Table 2-13 Coatings Performance Improvement with Different Paint Systems

Paint System	Performance Improvement-S-104 over Control
Acrylic	
White gloss spray paint (Rohm and Haas WS-24)	Improved coverage on metal-oily surface, significant reduction of flash rusting, reduction of foam-improved leveling and brushability
Green Semigloss (Rhoplex AC-658)	
Polyester-alkyd	
Black dip primer Ashland formulatory	Improved coverage and flow-faster drying times-improved foam control on oily and clean surfaces
Epoxy	
Industrial bake type	Improved wettability on oily substrates, improved coatability application, reduced foam levels
Vinyl acetate	
Interior-exterior paint Airflex 728 emulsion	Improved pigment wetting and color development. Enhanced grind efficiency, reduced foam levels

2.16.7 Wetting and Metal Applications: Surfynol Surfactants

The wetting speed in the Surfynol 400 series is greatest at 40 to 50% ethylene oxide content, as shown in Fig. 2-9. The Draves wetting times (cotton) for S-104 and S-440 are essentially identical and are shown as one curve. The steepness of the Surfynol wetting curve and the attainment of instantaneous wetting (zero time) illustrates the unique wetting speed of these surfactants [167,173]. The uniqueness of this wetting can be further shown by making the Draves wetting test more sensitive. By decreasing the weight of the standard 3.0-g hook attached to the cotton skein by a factor of 100, Surfynol 450 surfactant, at a concentration of 0.20%,

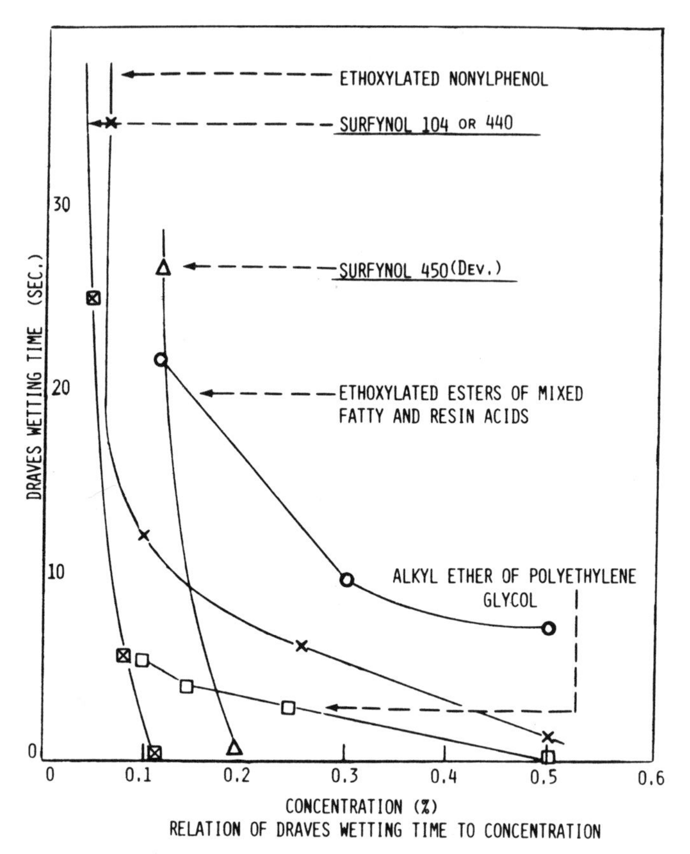

Figure 2-9 Comparison of surfactant wetting.

still gives instantaneous wetting, while conventional nonionic ethoxylates show little or no wetting (Draves time, 100-300 sec).

Ethoxylated Surfynol products (S-400 series), particularly S-465, are corrosion inhibitors for steel in dilute (5%) hydrochloric, phosphoric, or acetic acids [173]. Corrosion inhibition performance of 90% or better at 77°F have been obtained for S-465 at levels of 0.1 to 0.2% in hydrochloric acid and as low as 0.02% in phosphoric or acetic acids. The use of Surfynol products 465 and 61 in mixtures gives average corrosion protection of 87 to 99% at the above concentrations, with synergistic performance also noted. At higher levels, Surfynol 400 surfactants can be used as steel pickling inhibitors in 2 N H_2SO_4 at temperatures of 100 to 180°F to give average protection of 80 to 98% [174], depending on temperature and inhibitor concentration. Acetylenic diols such as dimethyl hexyndiol and dimethyl octyndiol (Surfynol 82) are also effective as pickling inhibitors. Both acetylenic alcohols and diols are highly effective in the related application, the prevention of hydrogen embrittlement of steels in acidic media.

The Surfynol 400 series also has the potential for use with acidic or temperature-unstable coatings, leading to acid formation via polymer degradation. It could be of use in PVC coatings to prevent both resin degradation and roller corrosion. The use of Surfynol products has also been recommended for PVC powder coatings for lowering viscosity and orange peel effect. Acetylenic diols are also used commercially in wire-drawing and metal-drawing operations, where ease of operation and a better quality metal surface are obtained.

2.16.8 Surfynol Products as Agricultural Adjuvants

Surfynol (S-) surfactants such as S-104, S-104-A, S-104-ES-75, S-TGE, S-400 seires, and S-82 are useful as both formulation aids and adjuvants in agricultural and related applications [173,174]. Varied properties such as high wetting speed, low foam, corrosion inhibition capability, dispersibility, and antistatic performance make them well suited for use in plant, animal, and poultry pesticides in either liquid concentrates or wettable powders.

These products are used commercially to promote the rapid dispersion of wettable powders (prevention of agglomeration and antistatic effects) and to improve the wetting and spreading action of pesticide solutions on plant surfaces for the control of weeds without harm to crops. Surfynol surfactants are also used in animal and poultry dip tanks for increased penetration of pesticide solutions through natural oil barriers on feathers and hair and to prevent excessive foam in the dip tank. Surfynol products 82, 104, 104-A, TGE, and the 400 series are listed as inerts in the Federal Register under 40 CFR, 180.1001 regulations (c) and (e). They are EPA-approved surfactants for use on crops, animals, and poultry.

Extensive university and industrial testing of Surfynol surfactants [175,176] indicate they have potential as tank mix additives for increasing the activity of herbicides in the control of weeds while still maintaining safety to crops. They can be used in such applications as preplant, preemergence, post, and no-till treatments for the control of weeds. University tests have shown them to have no harmful phytotoxic effects on crops such as corn, soybeans, cotton, spring wheat, rice, tomatoes, lettuce, etc. When used in concentrations of 0.1 to 1.0% of the spray at rates of 20 to 60 gal/acre, they are superior to, or at the minimum equivalent to, commercial adjuvants or surfactants. Both greenhouse [175] and university field tests [173,176] have supported the effectiveness of Surfynol adjuvants used on a variety of weeds and crops.

Surfynol adjuvants, particularly S-104, have been patented [177] as a novel composition, in combination with the herbicide Avenge (1,2-dialkyl-3,5-diphenylpyrazolium salt), for selective control of wild oats in spring wheat production. In this application, Surfynol 104 surfactant is claimed to function as a safening agent, minimizing phytotoxic damage to the wheat caused by the herbicide. University field tests [178] with corn and soybean herbicides have shown that Surfynol adjuvants are effective in lowering the rate at which herbicides are usually recommended compared with their recommended rate. The applications are for control of annual and perennial weeds in such crops as corn, soybeans, dry beans, barley, and spring wheat.

Typical of the more important EPA-approved Surfynol adjuvants are Surfynol products 82, 104, SE, TGE, Surfynol 400 series, and Blend C (S-400 series + emulsifier). Surfynol adjuvants are listed as inerts in Environmental Protection Agency (EPA) regulation 40 CFR 180.1001 paragraphs (c) and (e), and are exempt from the requirement of tolerance. They are consequently approved for use without limitation on harvested food crops (up to fruit appearance) and in animal and fowl pesticide formulations.

The superior performance of Surfynol adjuvants in herbicide-weed control studies is shown below [176]:

1. The rate of Atrazine used in early posttreatments for weed control in sweet corn can be lowered to 0.5 lb/acre active ingredient (aia) from its usual level of 1.5 to 2.0 lb aia if a Surfynol 400 adjuvant formulation (blend C) with an agricultural oil (11 N) is employed at 0.5% (vol %) of the spray. The adjuvant formulation comprised S-440, S-465, and emulsifier (80% Surfynol) used with 15% 11 N oil.
2. Surfynol products 465, SE, TGE, and blend C, used in posttreatments for soybean production at 0.1, 0.5, and 1.0% of the spray, gave superior yields (15-20 bu/acre) versus the control at 13 to 14 bu/acre. The herbicide Chloroxuron (Tenoran 50W) could be used with Surfynol adjuvants at a 50% rate reduction (1.0 lb aia) while still giving superior weed control versus standard Surfynol-free treatments at 1.5 lb aia.

3. Technical grades of both Atrazine and Alachlor, containing no inerts, when used alone or in mixtures of both herbicides with a variety of Surfynol adjuvants, gave excellent weed control (90-100% kill) with no adverse effects to corn. Both pre- and early posttreatments were used in these tests.
4. Surfynol surfactants used at 0.5% of the spray with trifluralin (Treflan 4 EC) at reduced rates (0.5-0.75 lb aia) in early posttreatments gave increased activity in the control of barnyard grass (Japanese millet) with minimum damage to corn and insignificant damage to beans. Trifluralin at present is used exclusively as a preemergence herbicide. It is recommended for use primarily with soybeans and vegetable crops, and not corn. This foliar application of trifluralin in early posttreatments may actually involve, at low rates of application, a combined foliar and soil incorporation. It has potential of being a new application for trifluralin.

Both greenhouse and field studies have supported the effectiveness of Surfynol adjuvants as tank-mix additives. Although the safening aspects are noted only in specific cases [177,179] they present an interesting potential for these unique adjuvants yet to be fully explored. Screening studies by Klepper [179] (University of Nebraska) first predicted the possible safening action of Surfynol 104 [177], while studying the adjuvant effects of various surfactants with herbicides, in enhancing nitrite accumulation (plant poisoning) in weed-leaf specimens.

2.16.9 Summary of Applications for Surfynol Products

The end-use applications and markets served by the Surfynol products are summarized in Table 2-14. It is apparent that the coatings area is important, particularly in metal-oriented applications and flow coatings. In the trade sales area, Surfynol products S-Tg, S-465, and S-485 confer such properties as freeze-thaw stability, increased color acceptance, hiding, and grinding efficiency for latex and related coatings. They are useful for vinyl acetate, acrylic, and butadiene-styrene latex systems. The use of S-104 and S-SE in water-based latices for improving coating performance on oily or dirty surfaces (particularly metal) is an important emerging use.

The agricultural market for Surfynol products at present is in an early growth stage and is regarded as having great potential.

Table 2-14 End-Use Applications of the Surfynol Products

Markets	Surfynol Products	End Products and Uses
Industrial coatings	S-104, S-104-A, S-104-E, S-104-H, S-TG	Water-based industrial primers and finishes; metal primer and finishing coatings; rinse aid for finishes; roll, spray, flow, and dip applications; pigment grinding, dispersing, and color development for lattices
Trade Sales	S-TG, S-104-E, S-465, S-485	Conventional latex building coatings for consumer and contractor application
Printing	S-104-E, S-TG, S-82, S-440	Water-based packaging inks, reproduction processes
Dyestuffs	S-TG, S-104, S-104-E	Optical brighteners, azo and azo disperse dyes, miscellaneous dye products
Agriculture	S-82, S-104, S-TG-E, S-104-A, S-465, S-SE	Formulation aids, defoamers, wettable powders, pesticide dip tanks, adjuvants
Adhesives	S-TG, S-104-H	Carpet, plywood, and varied industrial adhesive products
Process		Defoaming and wetting aid, petroleun and chemical processes

Surfynol products are being used as formulation aids for crop herbicides and for new biological control agents. Applications include wettable powders, emulsifiable concentrates, flowable concentrates, and adjuvant formulations for use with herbicides and other pesticides. The ability to complex and wet with great effectiveness makes these products attractive for plant and animal pesticide applications.

2.17 ACETYLENIC CHEMICALS AND DERIVATIVES: OVERVIEW

Reppe chemicals, dating back to the 1940s in Germany and the mid-1950s in the United States, have dominated and continue to dominate both production and dollar volume of acetylenic chemicals and their derivatives. In 1956 GAF, then known as General Aniline and Film Corporation and controlled by the U.S. government under the Alien Custodian Act, as a result of the defeat of Germany in World War II, built its first Reppe plant at Calvert City, Kentucky. This plant utilized calcium carbide acetylene, supplied "across the fence" from the AIRCO facility. In 1958 AIRCO also completed a plant for the production of acetylenic alcohols and diols at the same location to supplement their smaller production facility at Middlesex, New Jersey.

In 1963 GAF further diversified its position in Reppe chemicals by adding a facility for the production of vinyl ethers at Calvert City. However, by 1968 the growth of its Reppe line required GAF to build a new plant at Texas City, Texas due to favorable Gulf Coast economics for petrochemical acetylene and formaldehyde supply, together with excellent port facilities for its growing foreign business. Since their completion, both plants have undergone several expansions, attesting to the excellent growth in this product line. In Secs. 2.1 to 2.12 the technology and uses for these acetylenics and their derivatives is discussed. It is likely that acetylenic chemicals and derivatives will grow at an overall rate in excess of 10%/yr in the years ahead.

2.17.1 1,4-Butanediol (BDO)

This Reppe derivative has emerged in recent years as the most important of the Reppe chemicals, based on its uses in the manufacture of tetrahydrofuran (THF), as a chain extender for polyurethanes and, more recently, for the production of polybutylene terephthalate (PBT) [180]. This thermoplastic polyester is considered to be one of the most important engineering plastics to be developed in recent years, since it possesses an excellent balance of physical properties and exceptional moldability [181]. Its growth rate from 1974 to 1977 has been estimated at 40%/yr.

Below are summarized the producers of butanediol both in the United States and abroad [181]. Besides the well-established Reppe route, other processes based on dichlorobutene, maleic anhydride, or newer technology are currently used in Japan. Although Reppe technology is predominant in the United States, alternative non-acetylene-based processes have been developed, and are discussed in Sec. 2.7.2.

Producer	Process	Capacity (t/yr)	Planned Expansion (t/yr)
GAF (U.S.A.)	Reppe	35,000	35,000
Du Pont (U.S.A.)	Reppe	50,000	?
BASF-Wyandotte (U.S.A.)	Reppe	?	25,000
BASF (West Germany)		70,000	40,000
GAF-Huels (West Germany)	Reppe	?	25,000
BASF (Ludwigs-haven, West Germany)	Reppe	?	25,000
Toyosoda (Japan)		6,000	
Hokkaido Yuki (Japan)		2,400	
Mitsubishi (Japan)		?	15,000

In the United States the use of butanediol is expected to grow from its 1974 volume of 65 million lb to 120 million lb by 1979 [180]. About half of this growth will be due to PBT resins. Acetylene usage for butanediol is expected to grow from its 1974 volume of 21 million lb to about 38 million lb by 1979. Production of butanediol is expected to increase to about 310 million lb [183] annually by 1982, accounting for the use of about 100 million lb of acetylene.

2.17.2 Tetrahydrofuran (THF)

Du Pont is the predominant producer of THF in the United States, utilizing the classical Reppe route described in Sec. 2.8. Its original plant capacity of 50 million lb was increased to 90 million lb in the period 1973-1974. Du Pont production in 1974 was estimated at 80 million lb, which in turn used 34 million lb of acetylene. It is estimated that acetylene usage will increase to 47 million lb by 1979, which will account for 112 million lb of THF (an average growth rate of 7% [180]. THF will continue to be used in the applications described in Secs. 2.8 and 2.8.2 to 2.8.4. Since the Du Pont THF process involves the intermediate formation of 1,4-butanediol, Du Pont, if it so chooses, can compete in the growing

PBT and related plastics markets, as well as having captive butanediol for use in its own plastics operations.

2.17.3 Other Reppe Chemicals

Below are summarized the more important Reppe chemicals manufactured by the General Aniline and Film Corporation (GAF) in the United States, together with estimated consumption figures [182]. THF, mainly a Du Pont product, is shown for comparison.

Propargyl alcohol is sold in small amounts as a corrosion inhibitor for oil-well acidizing and steel-mill pickling, in addition to its use as a chemical intermediate. Its sales in the United States may be about 2 million lb annually. However, most of its production, as a coproduct of the ethynylation of formaldehyde, is recycled to produce additional butyndiol.

Product	1974 Consumption (MM lb)
1,4-Butanediol (total usage)	65
(a) Captive usage	35
(b) Merchant sales	30
Tetrahydrofuran (THF)	>80
γ-Butyrolactone	28-32
2-Pyrrolidone	13-17
N-Vinyl-2-pyrrolidone	10-12
Vinyl ethers	10-15
Propargyl alcohol (1-propyn-3-ol)	?
2-Butyn-1,4-diol	?
2-Buten-1,4-diol and other derivatives	6-8
N-Methyl-2-pyrrolidone	10-12

Most of the butyndiol produced in the United States or abroad is converted to the Reppe derivatives shown above (see Secs. 2.5-2.12). However, some butyndiol is sold as an intermediate for the manufacture of specialty products or as a metal treatment agent (corrosion inhibitor, pickling aid, electroplating additive). It is doubtful if these uses are in excess of 3 million lb. The principal use for γ-butyrolactone (shown above) is as an intermediate for 2-pyrrolidone and N-methyl-2-pyrrolidone. Probably the most important Reppe product in the GAF line is N-vinyl-2-pyrrolidone, since, through its polymer and copolymer derivatives, it is used in many applications with excellent growth potential (see Secs. 2.11-2.12.4).

2.17.4 Miscellaneous Acetylenic Specialty Chemicals

Originally, Air Reduction (up to 1971), and now Air Products and Chemicals, produced miscellaneous acetylenic alcohols and diols and derivatives made from them. These products, described in Secs. 2.15 to 2.16.9, are lower in production volume than their Reppe counterparts and are produced either at Calvert City, Kentucky or at Middlesex, New Jersey. The Calvert City plant has been expanded twice since the original 2 million lb/yr plant was built by Air Reduction in 1958. The Surfynol line of products is important in such industrial uses as coatings, metal treatments, pesticides, and general wetting-defoaming applications. The entire product line enjoys diversified uses in such areas as vitamins A and E, flavors and fragrances, corrosion inhibitors, stabilizers for chlorinated solvents, pesticides, and drugs. Because of the reactive nature of these acetylenic compounds, they are of great interest as building blocks for new compounds with unique applications.

REFERENCES

1. J. W. Reppe, O. Schlichting, K. Klager and T. Toepel, *Justus Liebigs Ann. Chem.*, *596*, 1-224 (1955).
2. J. W. Reppe, O. Schlichting, K. Klager and T. Toepel, *Justus Liebigs Ann. Chem.*, *601*, 81-138 (1956).
3. J. W. Reppe, *Acetylene Chemistry*. University Microfilms, Ann Arbor, Michigan (1949); PB Report 18852-S (trans. from the German), U.S. Dept. of Commerce, Washington, D.C. (1945).
4. J. W. Copenhaver and M. H. Bigelow, *Acetylene and Carbon Monoxide Chemistry*, Reinhold, New York, 1949; University Microfilms, Ann Arbor, Michigan (1949).
5. M. E. Chiddix, *Ann. N.Y. Acad. Sci.* *145*, 154-154 (1967).
6. E. V. Hort, U.S. Patent 3,920,759 (November 18, 1975).
7. G. L. Moore, U.S. Patent 3,218,362 (November 16, 1965).
8. C. McKinley, F. Fahnoe, and D. L. Fuller, U.S. Patent 2,712,560 (August 2, 1955).
9. R. J. Tedeschi, H. C. McMahon and M. S. Pawlak, *Ann. N.Y. Acad. Sci.* *145* (1), 93-100, 106 (1967).
10. R. J. Tedeschi, A. W. Casey, G. S. Clark, R. W. Huckel, L. M. Kindley and J. P. Russel, *J. Org. Chem.* *28*, 1740-1743, 2480 (1963).
11. C. J. S. Appleyard and J. F. C. Gartshore, BIOS Report 367; U.S. Dept. Commerce, OTS Report, PB 28556, Washington, D.C. (1945).

12. D. L. Fuller, FIAT Report 926; OTS Report, PB 80334, U.S. Dept. Commerce, Washington, D.C. (1946).
13. W. E. Hanford and D. L. Fuller, *Ind. Eng. Chem. 40,* 1171 (1948).
14. J. W. Reppe and E. Keyssner, U.S. Patent 2,232,867 (February 25, 1941).
15. E. Keyssner and E. Eichler, U.S. Patent 2,238,471 (April 15, 1941).
16. J. W. Reppe, A. Steinhofer, H. Spaenig and K. Locker, U.S. Patent 2,300,969 (November 3, 1942).
17. F. Fahnoe, U.S. Patent 2,527,358 (October 24, 1950).
18. *Chem. Week,* Another butanediol entry, April 10, 1974, p.16.
19. R. E. Kirk and D. E. Othmer, *Encyclopedia of Chemical Technology,* Suppl. 2., Interscience, New York, 1960, pp. 36-64 and references cited therein.
20. R. J. Tedeschi and G. S. Clark, U.S. Patent 3,462,499 (August 19, 1969).
21. General Aniline and Film Corp. (GAF), Propargyl alcohol, *New Products Bulletin,* No. A-103 (1952).
22. General Aniline and Film Corp. (GAF), Propargyl alcohol, *New Products Bulletin,* No. AP-59-2.
23. General Aniline and Film Corp. (GAF), 2-Butyne-1,4-diol, *New Product Bulletin,* No. 5M-7-62.
24. General Aniline and Film Corp. (GAF), *Technical Bulletins* No. 7543-029; 7543-037; 7543-059; 7543-077; 7543-113; 7543-134; 7543-144; 7543-200.
25. H. Berger and R. Linke, U.S. Patent 2,603,622 (July 15, 1952.
26. H. Iserson, U.S. Patent 2,801,160 (July 30, 1957).
27. F. A Hessel, U.S. Patent 2,946,716 (July 26, 1960).
28. T. R. Hopkins, R. P. Neighbors, P. D. Strickler and L. V. Phillips, *J. Org. Chem., 24,* 2040 (1959); See also *Chem. Eng. News,* November 2, 1959, p. 60; *Chem. Week,* December 13, 1958, p. 82.
29. G. Schultze, German Patent 821,984 (November 22, 1951.
30. H. Hopff and C. W. Rautenstrauch, U.S. Patent 2,304,917 (December 15, 1942).
31. E. Muller, O. Bayer, S. Peterson, H. F. Piepenbrink, H. F. Schmidt and F. E. Weinbrenner, *Rubber Chem. Technol. 26,* 493 (1953).
32. C. S. Marvel and J. H. Johnson, *J. Am. Chem. Soc.* 72, 1674 (1950).
33. R. Romanet, *Compt. Rend.,* 236, 1044-1046, 1176 (1953).
34. P. N. Rylander, *Catalytic Hydrogenation Over Platinum Metals,* Academic, New York, 1967, pp. 59-78.

35. C. W. Bird, *Transition Metal Intermediates In Organic Synthesis*, Academic, New York, 1966, pp. 248-271.
36. Br. Patent 772,212 (April 10, 1957).
37. H. Schwartz, U.S. Patent 2,840,598 (June 24, 1958).
38. O. Kardos, U.S. Patent 2,849,353 (August 26, 1958).
39. A. P. Bonnani, *Space/Aeronaurics 35,* 95 (1961).
40. *Europ. Chem. News*, Du Pont chooses acetylene route to tetrahydrofuran, December 16, 1966, p. 34.
41. *Chemical Economics Handbook*, 605.5022, *O-R* (August 1957), Stanford Research Institute (SRI), Menlo Park, Calif., 1957.
42. Mitsubishi hydrogenation Process, W. German Offenleg 2,144,316 (March 16, 1972).
43. A. M. Brownstein and H. L. List, *Hydrocarb. Proc. 9,* 159-162 (1977).
44. C. C. Cumbo and K. K. Bhatia, U.S. Patent 3,929,915 (December 30, 1975).
45. Mitsubishi Chemical Patents (Butanediol), W. German Offenleg 2,345,160 (March 24, 1974); 2,424,539 (December 12, 1974).
46. Mitsubishi Chemical Patents (Butanediol), W. German Offenleg 2,504,637 (August 14, 1975); 2,505,749 (August 14, 1975); 2,510,088 (September 18, 1975); 2,510,089 (September 18, 1975).
47. BASF Patent, W. German Offenleg 2,444,004 (March 25, 1976).
48. BASF Patent, W. German Offenleg, 2,454,768 (May 26, 1976).
49. I. E. Du Pont de Nemours & Co., Recovery of THF booklet, New York.
50. *Chemical Economics Handbook, Formaldehyde,* (b. Tetrahydrofuran), 685.50333 *K* (August 1970), Stanford Research Institute (SRI), Menlo Park, Calif., 1970.
51. M. G. Erskine, *Chemical Economics Handbook, Furfural,* 660.5020 *C* (April 1968), Stanford Research Institute (SRI), Menlo Park, Calif., 1968.
52. *Oil, Paint, Drug Rep.*, Furfural, July 15, 1968, p. 27.
53. *Oil, Paint, Drug Rep.*, Petro-Tex has a process to make THF via butadiene, April 29, 1968, p. 49; July 15, 1968, p. 27.
54. W. F. Brill and A. J. Besozzi, U.S. Patent 3,238,225 (March 1, 1966).
55. J. Kanetaka, T. Asano and S. Masamune, *Ind. Eng. Chem. 62,* (4), 24-32 (1970).

56. *Chem. Week*, Joining the petrochemical parade, March 15, 1968, pp. 63, 66.
57. T. Yoshimura, *Chem. Eng. 76,* (17) August 11, 1969, pp. 70-72.
58. J. L. Blackford, *Chemical Economics Handbook, Tetraethyl Lead and Tetramethyl Lead,* 671.5040, *A (October 1968),* Stanford Research Institute (SRI), Menlo Park, Calif., 1968.
59. *Chem. Week*, What will it cost to get the lead out?, October 13, 1971, p. 59.
60. *Chem. Eng. News*, Phasing lead out of gasoline, February 6, 1978, pp. 12-16.
61. I. E. Du Pont de Nemours & Co., Du Pont Tetrahydrofuran As a Chemical Intermediate (booklet), New York.
62. I. E. Du Pont de Nemours & Co., THF, Properties and Uses, New York.
63. I. E. Du Pont de Nemours & Co., THF As a Reaction Solvent, New York.
64. P. F. Lewis, *Chemical Economics Handbook, Elastomer Fibers (Spandex Fibers),* 543.3720 *A-K (August 1966),* Stanford Research Institute (SRI), Menlo Park, Calif., 1966.
65. E. O. Langerak, L. J. Prucino and W. R. Remington, U.S. Patent 2,692,873 (October 26, 1954); 2,692,874 (October 26, 1954).
66. F. S. Martin, U.S. Patent 2,751,363 (June 19, 1956).
67. Gamma Butyrolactone, PB Report 60902, U.S. Dept. Commerce, Washington, D.C.
68. General Aniline and Film Corp. (GAF), Butyrolactone, *Technical Bulletin* 2M-1-64, New York.
69. P. F. Lewis, *Chemical Economics Handbook, Polyamide Fibers (Nylon Fibers),* 534.4120 *A (December 1968),* Stanford Research Institute (SRI), Menlo Park, Calif., 1968.
70. W. O. Ney, U.S. Patent 2,638,463 (May 12, 1953).
71. W. O. Ney, U.S. Patent 2,739,959 (March 27, 1956).
72. General Aniline and Film Corp. (GAF), M-Pyrol, *Technical Bulletins* 7543-026, 7543-038, 7543-118; CHEMFO #73.
73. U. Wagner and H. M. Weitz, *Ind. Eng. Chem. 62,* (4), 43-48 (1970).
74. S. A. Miller, *Acetylene, Vol. 2,* Academic, New York, 1966, pp. 338-340.
75. PB Report 67620, U.S. Dept. Commerce, Washington, D.C., 1945.
76. R.A. Labine, *Chem. Eng. 67,*(4), 341-351 (1960).

77. General Aniline and Film Corp. (GAF), PVP-Preparation, Properties, Applications in the Blood Field and Other Branches of Medicine, New York (1951), pp. 1-174.
78. General Aniline and Film Corp. (GAF), Plasdone, AP-123, New York.
79. *Adhesives Age*, PVP and modified vinylpyrrolidone resins, *January 1969*, pp. 52-56.
80. H. B. Kellogg, PVP-Iodine in agricultural pest control, *Farm Chem. 119*, 41 (1956).
81. R. A. Clemens and A. J. Martinelli, PVP in the clarification of wines and juices, *Wines and Vines 39*, 55 (1958).
82. Printing and Surface Coating Art, U.S. Patent 2,944,912 (July 12, 1960); 2,978,428 (April 4, 1961); 2,719,831 (October 4, 1955).
83. F. J. Prescott, E. Hahnel, and D. Day, Cosmetic PVP, *Drug. Cos. Ind. 93*, 443-5, 540-1, 629-30, 702, 739 (1963).
84. T. Cifelli, PVP aerosol spray patent, *Drug Cos. Ind. 90* (4), 467 (1962).
85. H. Goldschmiedt, Soap Sheets, *Soap Chem. Spec. 33*, 47 (1957).
86. E. C. Hansen, C. A. Bergman and D. B. Witwer, PVP- A versatile compound, *Proc. Am. Assn. Text. Chem. Col., Am. Dyestuff Reptr. 43*, 72, (1954).
87. T. P. Callenan, Powdered resin Improves mechanical properties of inorganic specialty papers, *M. R. L. Progress*, September 1955, pp. 1-6; U.S. patents, 3,008,867, 3005,745; 2,901,390; 3,036,950; 3,081,219.
88. F. J. Prescott, Pharmaceutical dispersions, *Tex. J. Pharm. 4*, 300 (1963).
89. G. P. Lehrman and D. M. Skauers, A comparative study of PVP and other binding agents in tablet formulations, *Drug Stand. 26*, 170 (1958).
90. P. F. Lewis, *Chemical Economics Handbook, Synthetic Fibers Production*, 543.1130*B*; 534.3520 *A-T*, *(November 1969)*, Standard Research Institute (SRI), Menlo Park, Calif., 1969.
91. H. W. Coover, U.S. Patent 2,790,783 (April 30, 1957.
92. General Aniline and Film Corp. (GAF), Data Sheets 752-65, Commercial Development Dept.
93. General Aniline and Film Corp. (GAF), GAF Polymers as Protective Colloids in Emulsion Polymerization, CHEMFO #55.
94. L. Ross and R. J. Limon, Protective colloids, *Paint and Varnish Production*, December 1968, pp. 52-56.

95. *Chem. Eng.*, Flow Diagram of GAF Grasselli, New Jersey Pilot Plant, *59* (6), 176-179 (June 1951).
96. K. Herrie, U.S. Patent 4,053,696 (November 17, 1975).
97. J. W. Reppe, U.S. Patent 1,959,927 (May 22, 1934).
98. J. F. Vitcha and J. R. Russel, Unpublished work (1960), Air Reduction Co., Murray Hill, N.J.
99. V. A. Sims and J. F. Vitcha, *I&EC Product Res. Dev. 2*, 293 (1963).
100. W. Watanabe and L. Conlon, *J. Am. Chem. Soc. 79*, 2828 (1957); U.S. Patent 2,760,990 (August 28, 1956).
101. J. R. Russel and M. W. Leeds, Unpublished work, Air Reduction Co., Murray Hill, N.J., 1967; U.S. Patent pending.
102. S. A. Miller, *Ethylene and Its Industrial Derivatives*, Ernest Benn, London, 1969; *Europ. Chem. News 7* (165), 25 (1965).
103. French Patent 1,423,314 (1966).
104. General Aniline and Film Corp. (GAF), Methyl Vinyl Ether, *Technical Bulletins* TA-77-2 (4M-4-63); Vinyl ethers 7543-007.
105. General Aniline and Film Corp. (GAF), Alkyl vinyl ether monomers, Preliminary data sheet, January 7, 1964.
106. General Aniline and Film Corp. (GAF), Copolymerization with methyl vinyl ether, Preliminary data sheet No. 336-64.
107. General Aniline and Film Corp. (GAF), Cetyl vinyl ether polymerization, Preliminary data sheet No. 346-65; Long-chain alkyl vinyl ethers, *Technical Bulletin* 7543-080.
108. Gantrez M, GAF registered trademark for Polyvinyl Methyl Ether, Brochure No. 326-63 (Polyvinyl methyl ether).
109. D. S. Otto, Vinyl Ether Modified Polymers (Technical Seminar), GAF Commercial Development Department, New York.
110. *Oil, Paint, Drug Rep.*, GAF adds capacity to copolymer line, August 26, 1968, pp. 5, 43 (cf. also May 1, 1968, p. 3).
111. *Oil, Paint, Drug Rep.*, Gantrez now commercial, February 21, 1966, p. 1.
112. Gantrez AN, GAF registered trademark for poly(methylvinylether) maleic anhydride copolymer; *Technical Bulletin 7543-017* and references cited, New York.
113. *Adhesive Age*, Wash-off labels foil switches, *12*(1), 34 (1969).
114. Gantrez VC, GAF registered trademark for alkyl vinyl ether copolymers with vinyl chloride, Commercial Development Products, New York.

115. R. J. Tedeschi, Personal observation.
116. Mercury Catalysts for Vinyl Fluoride; Patents: Br. Patent 479,421 (1938); U.S. 2,401,850 (1946); U.S. 2,479,957 (1949); Br. Patent 590,381 (1947); U.S. 2,519,199 (1950); Fr. Patent 1,324,408 (1963); U.S. 2,437,307 (1948); U.S. 2,480,021 (1949); Br. Patent 916,130 (1963).
117. Process Patents for 1,1-Difluoroethane: U.S. 2,425,991 (1947); U.S. 2,830,090 (1958); U.S. 3,073,871 (1963); Br. 896,459 (1962).
118. R. M. Hedrick, U.S. Patent 2,695,320 (November 23, 1954).
119. *Mod. Plastics*, PVF and acrylic films vie for paint markets, *46*, (9), 82 (1969).
120. R. E. Kirk and D. E. Othmer, *Encyclopedia of Chemical Technology, Vol. 9*, Interscience, New York, 1960, pp. 835-845 and references cited therein.
121. *Chem. Week*, Business newsletter, May 6, 1970.
122. *Chemical Economics Handbook*, Acetylene, 605.5020 *W*, (April 1968), (vinyl fluoride and vinylidene fluoride), Stanford Research Institute (SRI), Menlo Park, Calif., 1968.
123. Vinylidene Fluoride Patents; U.S. 2,551,573 (May 8, 1951); 2,628,989 (February 17, 1953); 2,774,799 (December 18, 1956).
124. *Building Feature*, Polyvinylfluoride for building finishes, July 1967, pp. 719, 721.
125. A. W. Johnson, *Acetylene Compounds, Vol. I, The Acetylenic Alcohols*, Edward Arnold Co., London, 1946, pp. 6-16, 28, 122, 137-143 (or University Microfilms, Ann Arbor).
126. R. A. Raphael, *Acetylenic Compounds in Organic Synthesis*, Academic, New York, 1955, pp. 1-15 and references cited therein.
127. T. F. Rutledge, *Acetylenic Compounds-Preparation and Substitution Reactions*, Reinhold, New York, 1968, pp. 146-244 and references cited therein.
128. H. G. Viehe, *Chemistry of Acetylenes*, Marcel Dekker, New York, 1969, pp. 207-245 and references cited therein.
129. R. J. Tedeschi, A. W. Casey and J. P. Russel, U.S. Patent 3,082,260 (March 19, 1963).
130. H. A. Stanbury and W. R. Proops, *J. Org. Chem. 27*, 279 (1962).
131. R. J. Tedeschi, U.S. Patent 3,663,628 (May 16, 1972).
132. R. J. Tedeschi, *J. Org. Chem. 30*, 3045 (1965).

133. R. J. Tedeschi, G. S. Clark, Jr., G. L. Moore, A. Halfon, and J. Improta, *I&EC, Proc. Des. Des. Dev.*, *7*, 303 (1968).
134. A. C. McKinnis, *Ind. Eng. Chem.* *47*, 850 (1955).
135. M. V. Stackelberg and H. R. Muller, *Z. Elektrochem.* *58*, 25 (1954).
136. S. A. Miller, *Acetylene, Vol. 1*, Academic, New York, 1965, pp. 74-116, 517-535.
137. A. Balducci and M. De Malde, U.S. Patent 3,283,014 (Nov. 1, 1966).
138. R. J. Tedeschi, U.S. and foreign patent applications, 1970, 1971; U.S. Patent 3,663,628 (May 16, 1972).
139. R. J. Tedeschi and G. L. Moore, *Ind. Eng. Chem. Prod. Res. Dev.*, *9*, 83 (1970).
140. A. Balducci and M. De Malde, Fr. Patent 1,230,067 (September 13, 1960); Belgian Patent 578,695 (1958); Italian Patent 7791 (1958).
141. V. Cariati, M. Massi, and A. Di Cio, U.S. Patent 3,301,771 (January 31, 1967).
142. A. Sturzenegger, U.S. Patent 3,496,240 (February 17, 1970).
143. R. J. Tedeschi, Unpublished work.
144. G. H. Whitfield, U.S. Patent 2,826,614 (March 11, 1958).
145. R. K. Frantz, U.S. Patent 3,105,098 (September 24, 1963).
146. H. Pasedach, Ger. Patent 1,643,710 (June 9, 1971).
147. J. J. Nedwick, *I&EC Proc. Des. Dev.*, *1*, 137 (1962).
148. N. Shachat and J. J. Bagnell, Jr., *J. Org. Chem.* *27*, 1498 (1962).
149. J. H. Blumenthal, U.S. Patent 2,996,552 (August 15, 1961).
150. Fr. Patent 1,573,026 (January 31, 1968); Ger. Patent 1,904,918 (June 9, 1971).
151. W. C. Meuly, *Amer. Perfumer Cosm.*, Synthetic terpene chemicals from isoprene, *85*(9), 123 (1970).
152. A. I. Nogaideli, *Zh. Prikl. Khim.* *38*, 1639 (1965); *Chem. Abstr.* *63*, 9798 (1965).
153. R. J. Tedeschi and G. L. Moore, *J. Org. Chem.* *34*, 435 (1969).
154. *Chem. Week*, New processes give isoprene more bounce, March 24, 1971, pp. 39-41.
155. Air Products and Chemicals, Inc., Methyl Butynol and Methyl Pentynol, *Technical Bulletin A-1*, Allentown, Pa.
156. R. J. Tedeschi, Personal estimate.

157. Air Products and Chemicals, Inc., Ethynyl Cyclohexanol, *Technical Bulletin A-3*, Allentown, Pa.
158. Air Products and Chemicals, Inc., Hexynol and Ethyl Octynol, *Technical Bulletin A-2*, Allentown, Pa.
159. Air Products and Chemicals, Inc., OW-1 Corrosion Inhibitor, *Technical Bulletin A-8*, Allentown, Pa.
160. Air Products and Chemicals, Inc., Tertiary Acetylenic Glycols, *Technical Bulletin G-1*, Allentown, Pa.
161. Air Products and Chemicals, Inc., Dimethyl Hexanediol, *Technical Bulletin G-2*, Allentown, Pa.
162. Air Products and Chemicals, Inc., Dimethyl Octanol (April 100), *Data Sheet A-4*, Allentown, Pa.
163. Air Products and Chemicals, Inc., Alkyl Acetylenes, *Technical Bulletin C-1*, Allentown, Pa.
164. K. Taguchi, A. Yamamoto and T. Ishihara, U.S. 4,026,954 (May 31, 1977).
165. Air Products and Chemicals, Inc., Surfynol 82 and Surfynol 104, Technical Bulletin S-1, Allentown, Pa.
166. Air Products and Chemicals, Inc., Surfynol 400 Series, *Technical Bulletin S-5*, Allentown, Pa.
167. M. W. Leeds, R. J. Tedeschi, S. J. Dumovich and A. W. Casey, *I&EC Prod. Res. Dev.*, *4*, 236 (1965).
168. Air Products and Chemicals, Inc., Surfynol 104 A and Surfynol 104 E, *Technical Bulletin S-3*, Allentown, Pa.
169. Air Products and Chemicals, Inc., Surfynol TG, *Technical Bulletin S-2*, Allentown, Pa.
170. Air Products and Chemicals, Inc., Surfynol PC Defoamer, *Technical Bulletin S-6*, Allentown, Pa.
171. Air Products and Chemicals, Inc., Surfynol 61, *Technical Bulletin S-4*, Allentown, Pa.
172. R. J. Tedeschi and V. A. Sims, Acetylenic Surfactants in coating and related applications, Air Products and Chemicals Inc., Allentown, Pa., unpublished report (1978).
173. R. J. Tedeschi and V. A. Sims, Acetylenic surfactants in agricultural applications; *Paper presented at the Sixteenth Annual Meeting of the Weed Science Society of America (WSSA)*, Denver, Col., February 2-5, 1976.
174. R. M. Hudson and K. J. Riedy, Metal Finishing, Westwood, N.J., 1964, pp. 3-7.
175. Greenhouse Study for Air Products and Chemicals, Inc., Screening of seven candidate surfactants with five herbicides, by Bio-Search and Development Co., Kansas City, Mo. (1977).

176. R. J. Tedeschi, University adjuvant studies, 1974-1977, Air Products and Chemicals, Inc., Allentown, Pa. (1977).
177. R. Demchak, U.S. Patent 4,011,062 (March 7, 1977).
178. R. Sweet, Weed Science Report, Vegetable Crops Department, Cornell University, Ithaca, New York (1977).
179. L. Klepper, Evaluation of surfynols for nitrite accumulation in leaf specimens, Department of Pathology, University of Nebraska, Lincoln, Neb.; unpublished report to Air Products and Chemicals, Inc., Allentown, Pa. (1975).
180. T. H. McGreevey, *Chemical Economics Handbook, Acetylene*, 605.5022 *O-S* (Acetylenic Chemicals), Stanford Research Institute (SRI), Menlo Park, Calif., 1975.
181. Y. Tsutsumi, Technological trends in 1,4-butanediol, *Chem. Econ. Eng. Rev.*, *8*, (5) (#95), 45-50 (1976).
182. *Chemical Economics Handbook, Acetylene*, 605.5022 *Q* (A. Reppe Chemicals), Stanford Research Institute, Menlo Park, Calif., 1975.
183. Koon-Ling Ring, *Chemical Economics Handbook, Olefins*, (Product review of acetylene), 300.5000 *N,O,P*, Stanford Research Institute, Menlo Park, Calif., 1978.

FIGURES AND TABLES

INDEX

A

M

N

O

P

T